Macmillan
Revised

ENCYCLOPEDIA
of SCIENCE

The Heavens

2

Stars, Galaxies, and the Solar System

Robin Kerrod

Macmillan Reference USA
New York

Published by:
Macmillan Library Reference
A Division of Macmillan Publishing USA
1633 Broadway, New York, NY 10019

Copyright © 1991, 1997 Andromeda Oxford Limited
An Andromeda Book
Devised and produced by Andromeda Oxford Ltd,
11-15 The Vineyard, Abingdon, Oxfordshire, OX14 3PX England

Macmillan edition copyright © 1991, 1997
Macmillan Reference USA

Library of Congress Cataloging-in-Publication Data

Macmillan encyclopedia of science. -- Rev. ed.
        p.   cm.
        Includes bibliographical references and index.
        Summary: An encyclopedia of science and technology, covering
such areas as the Earth, astronomy, plants and animals, medicine,
the environment, manufacturing, communication, and transportation.
    ISBN 0-02-864556-1 (set)
    1. Science--Encyclopedias, Juvenile. 2. Engineering-
-Encyclopedias, Juvenile. 3. Technology--Encyclopedias, Juvenile.
[1. Science--Encyclopedias. 2. Technology--Encyclopedias.]
I. Title: Encyclopedia of science.
0121 M27 1997
500--DC20   96-36597
                                                                  CIP
                                                                   AC

Volumes of the *Macmillan Encyclopedia of Science*
Set ISBN 0028645561
  1 *Matter and Energy*   ISBN 002864557X
  2 *The Heavens*   ISBN 0028645588
  3 *The Earth*   ISBN 0028645596
  4 *Life on Earth*   ISBN 002864560X
  5 *Plants and Animals*   ISBN 0028645618
  6 *Body and Health*   ISBN 0028645626
  7 *The Environment*   ISBN 0028645634
  8 *Industry*   ISBN 0028645642
  9 *Fuel and Power*   ISBN 0028645650
 10 *Transportation*   ISBN 0028645669
 11 *Communication*   ISBN 0028645677
 12 *Tools and Tomorrow*   ISBN 0028645685

Printed in the United States of America

Introduction

This volume surveys what astronomers have discovered about the universe and how it developed. Much of the information is presented in photographs, drawings, and diagrams. They are well worth your attention.

To learn about a specific topic, start by consulting the Index at the end of the book. You can find all the references in the encyclopedia to the topic by turning to the final Index, covering all 12 volumes, located in Volume 12.

If you come across an unfamiliar word while using this book, the Glossary may be of help. A list of key abbreviations can be found on page 87. If you want to learn more about the subjects covered in the book, the Further Reading section is a good place to begin.

Scientists tend to express measurements in units belonging to the "International System," which incorporates metric units. This encyclopedia accordingly uses metric units (with American equivalents also given in the main text). In illustrations, to save space, numbers are sometimes presented in "exponential" form, such as 10^9. More information on numbers is provided on page 87.

Contents

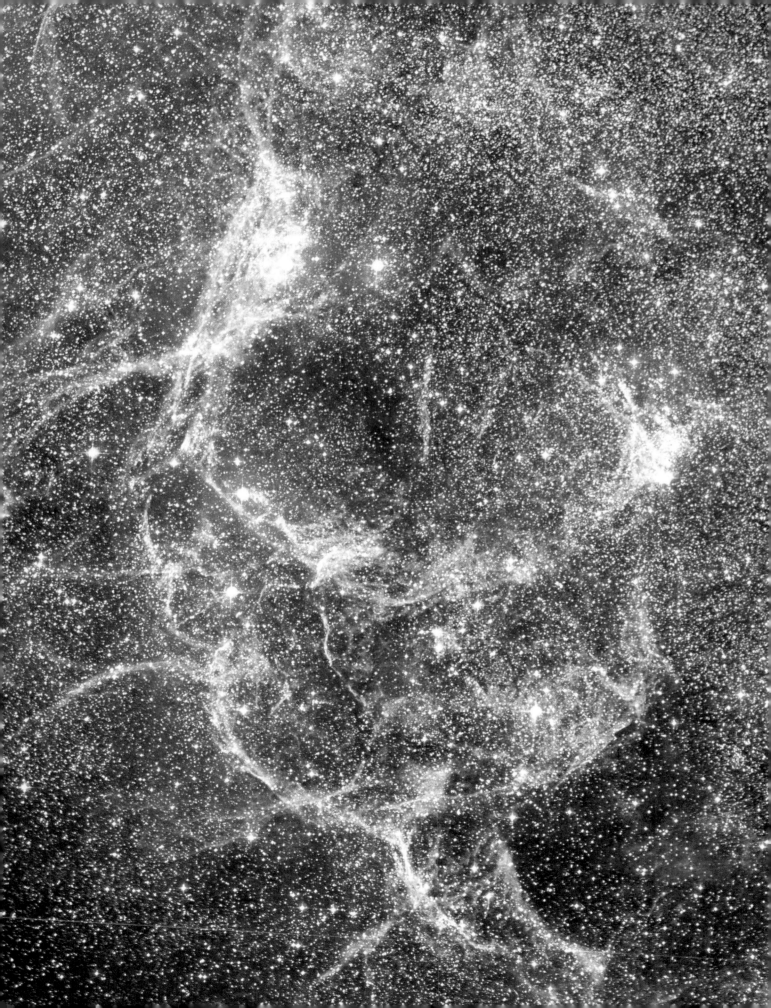

Part One

Stars and galaxies

When we gaze into the inky blackness of the night sky, we are looking deep into space, into the depths of the Universe. All things that exist – the Earth, Moon, Sun, planets, stars, galaxies, and even space itself – make up the Universe. We know much about the Universe and the bodies it contains. We think we know how it began and even when.

The science that studies the Universe is astronomy. It began in the Middle East at least 5,000 years ago, and so is one of the oldest of the sciences. Yet it is also one of the newest. Using the latest telescopes and Space Age techniques, astronomers are continually coming up with astonishing new discoveries, such as quasars and pulsars. And they are finding evidence of awesome bodies they call black holes, which gobble up everything, even light.

◄ The remains of a supernova in the constellation Vela. It has been expanding into space since the explosion of a star about 10,000 years ago. Only a small part is shown here.

Measuring the heavens

People began stargazing thousands of years ago, and the practice grew into the science of astronomy. Perhaps surprisingly, the heavens have changed little over this period. Early astronomers saw the same kind of constellations, or star patterns, as we do today. They thought that the stars were fixed to the inside of a great celestial sphere that spun around the Earth. They had no idea of how big the Universe was. Only in the twentieth century did astronomers gain an idea of the true scale of the Universe. It is bigger than anybody can imagine.

SPOT FACTS

• The Sun lies some 150 million km (93 million mi.) away from Earth. Its light takes 8.5 minutes to reach us, and 5.5 hours to reach Pluto, the farthest planet in the Solar System.

• A light year is a unit of measuring equivalent to the distance that light can travel in one year. One light years equals 9.5 million million kilometers (6 million million miles).

• Traveling at its usual speed in orbit, the space shuttle would take 100,000 years to reach the nearest star, Proxima Centauri, which is about 4.25 light years away.

• Light from the Large Magellanic Cloud, the nearest galaxy, takes 170,000 years to reach the Earth.

• The day's length is increasing by 1.7 thousandths of a second per century. Our days are 1/10th of a second longer than 6,000 years ago.

• Most star names are Arabic and derive from the star's position in the constellation. Hence Rigel in Orion comes from the Arabic for "Orion's left foot."

• Seen from space, there are about 6,000 stars visible in the sky brighter than magnitude 6 (see page 16). But from Earth on a clear dark night, you can see no more than 3,000 stars at a time.

• The faintest stars that can be detected by the Hubble Space Telescope are fainter than magnitude 30 (see page 16).

• The largest constellation is Hydra. It extends across more than a quarter of the sky.

• Two stars were named in the nineteenth century as "Sualocin" and "Rotanev." These are the Latin names of an Italian observatory assistant, Nicolaus Venator, spelled backward.

• The celestial poles trace a circle in the sky over 25,800 years because of a polar wobble called precession. The North Star will no longer be close to the pole in 2,000 years time.

• The ancient Greeks named 48 constellations in the Northern Hemisphere. Those added in the eighteenth century include Antlia, the Air Pump, and Caelum, the Chisel.

• The smallest constellation is Crux, the Southern Cross. Only 9 degrees across, it can be covered by the palm of your hand at arm's length. It appears on the flags of Australia and New Zealand.

• The technique known as parallax used to calculate the distance of stars is so precise that the thickness of a human hair can be detected at a distance of 1 km (0.62 mi.) away.

• The parallaxes of 100,000 stars have been measured by a satellite called *Hipparcos* with such accuracy that our knowledge of the distances of the stars has been increased 100 times.

• It was reported in 1996 that the nearest star to the Earth (apart from the Sun), Proxima Centauri, rotates on its axis about every 31 days. Its light varies as large sunspots – or "starspots" – and flares appear on the surface.

• The farthest object that we can see with the naked eye is the Andromeda galaxy. Its light takes more than 2 million years to reach the Earth.

Scale of the Universe

Simply by looking up at the night sky we can see that the Universe of stars and space is vast. But just how big is it? How far away are the stars and the galaxies we can see with our eyes and through telescopes?

Nobody had any real idea of the distances to the stars until 1838. In that year the German astronomer Friedrich Bessel used a method called parallax to measure the distance to a star in the constellation Cygnus. The distance turned out to be 105 trillion km (65 trillion mi.).

Light travels extremely quickly, at the fastest speed we know. Yet over such a vast distance, it takes the light from that star 11 years to reach us. Astronomers say that the star is 11 light-years away.

Other stars are up to tens of thousands of light-years away. The galaxies are even more remote. But the most remote objects of all appear to be the quasars. They lie up to 13 billion or more light-years away at the edge of the observable Universe.

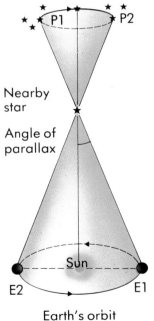

How far away?

◄ We can measure the distance to nearby stars by using the method of parallax. This works on the principle that a nearby object appears to change its position (P1/P2) when seen from different viewpoints. We can view a nearby star from opposite sides of the Earth's orbit (E1/E2). From each side, the star appears in a different position against the background of distant stars. From the amount the star appears to shift, its distance can be calculated. The distance from Earth at which the parallax of a star is an angle of 1 second (equal to 1/3,600th of a degree) is called a parsec.

► We can gain an idea of the vastness of the Universe by tracing how the Solar System fits into it, and looking at larger and larger areas of space. The Sun is part of a galaxy, which is part of a cluster of galaxies, which is part of a supercluster. Many superclusters make up the whole of the Universe.

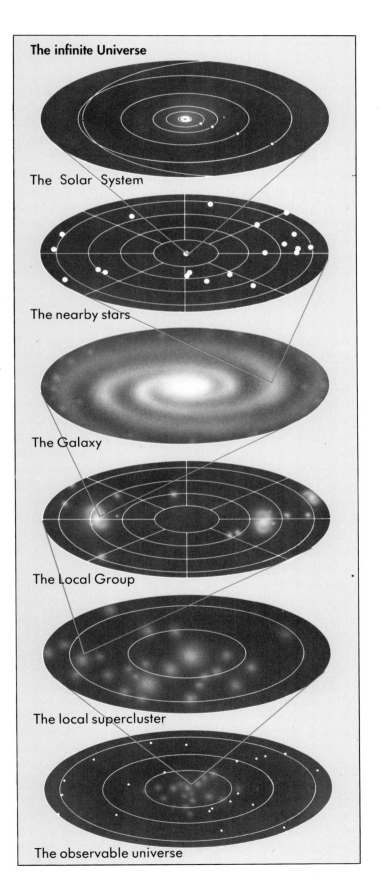

The infinite Universe

The Solar System

The nearby stars

The Galaxy

The Local Group

The local supercluster

The observable universe

The celestial sphere

When we look at the night sky for any length of time, the stars appear to wheel round the sky. It is as if they were fixed to the inside of a great globe, or sphere, that rotates around the Earth. Ancient astronomers believed there was such a sphere. But we now know that it does not exist, and that it is the Earth rather than the stars that rotates. Nevertheless, astronomers find the idea of a celestial sphere extremely useful for pinpointing the positions of the stars.

The Earth revolves on its axis once every 24 hours, relative to the Sun. This is the basis of our ordinary, solar time. But during the same period the Earth moves slightly along its orbit. And relative to the stars it takes only 23 hours 56 minutes to revolve once. This period is known as the sidereal day.

Astronomers use this period as the basis of star time, or sidereal time. Stars rise and set at the same sidereal time each day, and reach the

▲ In a long-exposure photograph, the stars appear to trail in arcs around the North Star, or polestar, because of the rotation of the Earth. The North Star is the white blob in the middle. At present it lies very close to the north celestial pole and hardly seems to move. Note the colors of the star trails.

▶ The Earth spins on its axis in space from west to east, making the stars appear to travel from east to west (1). Viewed from the Earth's North Pole, the stars appear to circle parallel with the horizon (2). Viewed from the Equator, the stars appear to rise and set vertically (3).

The whirling heavens

1 Because of the Earth's spin, the stars appear to move from East to West

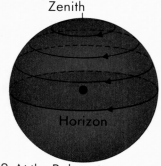

2 At the Poles, stars move parallel to the horizon

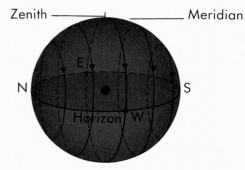

3 At the Equator, stars rise and set vertically

same points in the sky at the same sidereal time. So sidereal time provides a means of locating the stars on the celestial sphere.

The revolving celestial sphere

The celestial sphere rotates about the same axis as the Earth. The north and south celestial poles are located where the Earth's axis meets the sphere, vertically above the Earth's poles. The celestial equator is the circle where the plane of the Earth's Equator meets the sphere. The stars appear to travel in circles parallel with the celestial equator. During the year the Sun appears to travel around the sphere in a circle called the ecliptic. Its path crosses the celestial equator at two points – the equinoxes – the spring (vernal) equinox on about March 21 and the autumnal equinox on about September 23. On these dates throughout the world, day and night are of equal length.

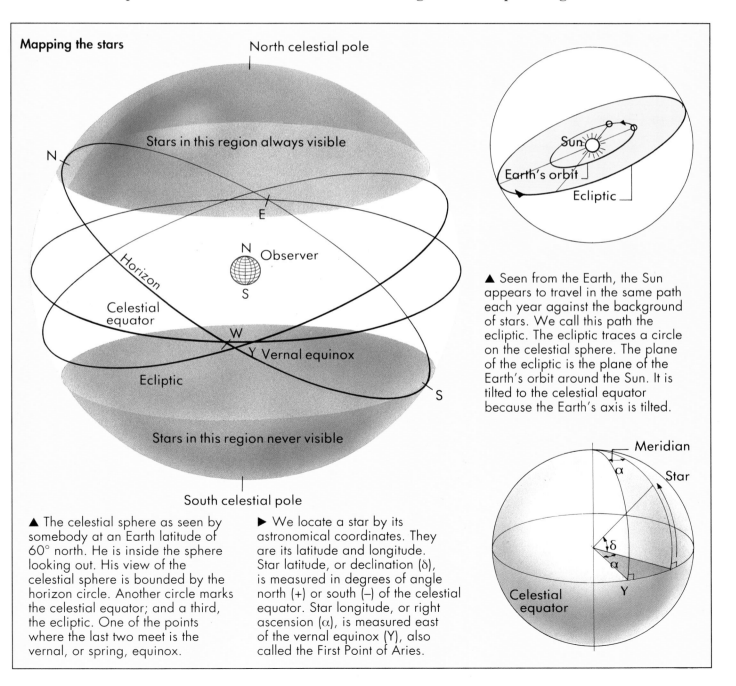

Mapping the stars

North celestial pole

Stars in this region always visible

N

E

N Observer
S

Horizon

Celestial equator

W

Y Vernal equinox

Ecliptic

S

Stars in this region never visible

South celestial pole

Sun

Earth's orbit

Ecliptic

▲ Seen from the Earth, the Sun appears to travel in the same path each year against the background of stars. We call this path the ecliptic. The ecliptic traces a circle on the celestial sphere. The plane of the ecliptic is the plane of the Earth's orbit around the Sun. It is tilted to the celestial equator because the Earth's axis is tilted.

Meridian

α

Star

δ

α

Celestial equator

Y

▲ The celestial sphere as seen by somebody at an Earth latitude of 60° north. He is inside the sphere looking out. His view of the celestial sphere is bounded by the horizon circle. Another circle marks the celestial equator; and a third, the ecliptic. One of the points where the last two meet is the vernal, or spring, equinox.

▶ We locate a star by its astronomical coordinates. They are its latitude and longitude. Star latitude, or declination (δ), is measured in degrees of angle north (+) or south (–) of the celestial equator. Star longitude, or right ascension (α), is measured east of the vernal equinox (Y), also called the First Point of Aries.

Constellations

Some of the brightest stars appear to make patterns in the sky, which we call constellations. The patterns remain much the same year after year, and change only very slowly. But usually the stars in a constellation are not close to each other in space. They appear that way because they lie in the same direction from us.

The earliest stargazers in ancient Babylon named the northern constellations they could see after figures they thought the star patterns looked like. The names passed to ancient Egypt, Greece, and Rome. We still use the Latin names for them. Some are animals: Ursa Major, the Great Bear; Cygnus, the Swan; and Leo, the Lion. Some are mythological people, such as Hercules, Orion, and Andromeda. Others are everyday objects, such as Crater, the Cup, and Libra, the Scales.

Some individual stars within a constellation have names of their own. For example, the brightest star in the constellation Canis Major, the Great Dog, is called Sirius. But in general astronomers identify a star in a constellation by a Greek letter, according to its brightness or position. The brightest is usually labeled alpha (α), the next brightest beta (β), and so on.

We measure the brightness of a star on a system pioneered by the ancient Greeks, who grouped stars they could see into six categories of brightness. The brightest were described as 1st magnitude and the dimmest as 6th magnitude. For exceptionally bright stars the scale is extended backward to give negative values. Sirius, for example, has a magnitude of -1.45. For stars too dim to see without a telescope, the scale is extended forward beyond 6.

◀ This illustration from an early book on astronomy shows constellations of the Northern Hemisphere. The artist has used a vivid imagination to create suitable figures to match the pattern of bright stars. The Great Bear (Ursa Major) stands out clearly, as it does in the heavens. A circle marks the ecliptic, and the constellations it goes through are known as the constellations of the zodiac. These play an important role in astrology.

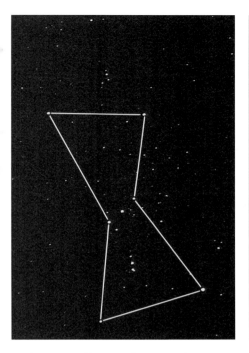

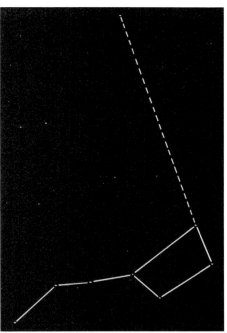

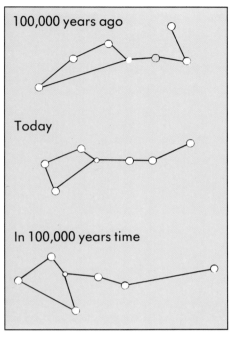

100,000 years ago

Today

In 100,000 years time

▲ The constellation of Orion, the Mighty Hunter, sits on the celestial equator and is easy to spot. The three stars across the middle form Orion's Belt.

▲ The Big Dipper is part of the constellation of the Great Bear (Ursa Major). The two stars at the front of the dipper's cup, pointing toward the North Star, are the Pointers.

▲ The seven main stars of the Big Dipper lie at different distances in space and move at different speeds. That is why the shape of the constellation gradually changes.

▲ An astrologer's star signs from the island of Bali. Astrologers believe that people's lives are affected by the positions of the planets among the constellations.

The main constellations

Latin name	Common name	Latin name	Common name
Andromeda		Dorado	Swordfish
Aquarius	Water Bearer	Draco	Dragon
Aquila	Eagle	Eridanus	River
Aries	Ram	Gemini	Twins
Auriga	Charioteer	Hercules	
Boötes	Herdsman	Hydra	Water Monster
Camelopardus	Giraffe	Leo	Lion
Cancer	Crab	Libra	Balance
Canes Venatici	Hunting Dogs	Lyra	Lyre
Canis Major	Greater Dog	Ophiuchus	Serpent Bearer
Canis Minor	Lesser Dog	Orion	
Capricornus	Sea Goat	Pegasus	Flying Horse
Carina	Keel	Perseus	
Cassiopeia		Pisces	Fishes
Centaurus	Centaur	Puppis	Stern
Cepheus		Sagitta	Arrow
Cetus	Whale	Sagittarius	Archer
Columba	Dove	Scorpius	Scorpion
Coma Berenices	Berenice's Hair	Serpens	Serpent
Corona Australis	Southern Crown	Sextans	Sextant
Corona Borealis	Northern Crown	Taurus	Bull
Corvus	Crow	Triangulum	Triangle
Crater	Cup	Ursa Major	Great Bear
Crux	Southern Cross	Ursa Minor	Little Bear
Cygnus	Swan	Vela	Sails
Delphinus	Dolphin	Virgo	Virgin

The starry heavens

The velvety black heavens are studded with stars of every description. There are massive red supergiants hundreds of times bigger than the Sun, and blazing hot blue stars that shine 10,000 times more brightly. There are stars that wink as regularly as clockwork, stars with companions, and stars that cluster together in their thousands. Astronomers today can tell us all manner of things about these stars – their size, mass, color, temperature, speed, and distance. All this, and more, they find out from faint smudges of starlight.

SPOT FACTS

• Over 5,000 stars can be seen with the naked eye.

• One of the most massive stars known is Plaskett's star in the constellation Monoceros, the Unicorn. It is a binary system (it travels with a companion star), and each star has more than 50 times the mass of the Sun.

• The most powerful star is the variable Eta Carinae, in the constellation Carina, the Keel. In the mid-1800s it became 4 million times brighter than the Sun.

• Barnard's star moves across the celestial sphere faster than any other star. Its movement (proper motion) is a little over 10 seconds of arc (1/360th of a degree) per year.

• The star Betelgeuse is one of the few that is large and close enough for details to be seen on it. But its disk as seen from Earth is no larger than a period on this page seen from a mile away.

• The most common stars in our Galaxy are dim red stars; 70 stars out of 100 are this type. Only about four in 100 are Sun-type stars.

• The star Alpha Centauri, the nearest bright star to the Sun, is so similar in brightness and color that from a distance they must appear as twin stars.

• Most variable stars take days to change brightness. But a faint star called CY Aquarii is so rapid that it doubles in brightness in 9 minutes.

• Of the 25 nearest stars to the Sun, only three are brighter.

• Measurements in 1995 and 1996 of small changes in the radial motions of some nearby stars suggested that they have planets going around them. These were the first detections of probable planets around ordinary stars other than the Sun.

• Some astronomers believe that life may have originated in giant molecular clouds in space, and that it spread to planets such as the Earth when the planets passed through the clouds.

• The star Algol in the constellation Perseus is a well-known eclipsing binary variable star (it travels with a companion star and fades every 2 days 21 hours). Its name comes from the Arabic word meaning "Ghoul" – suggesting that medieval astronomers were aware of its properties.

• The dramatic colors of nebulae in photographs are real, but are too faint to be seen even with large telescopes. The strong red of hydrogen is detected by film, but the eye is very insensitive to that color.

• Dust clouds in the constellation of Cygnus are known as the "Great Rift" because they give the appearance that the Milky Way is split in two. The dust clouds are easily visible with the naked eye on a clear night.

• No stars are strongly colored, but red giants such as Betelgeuse and Aldebaran are obviously orange. Their surface temperature is similar to that of a 40-watt light bulb.

• The discoverer of the variability of Delta Cephei was a young deaf and dumb amateur astronomer named John Goodricke, who lived in eighteenth-century York, England. He was awarded the highest scientific honors for his discovery.

Varieties of stars

All the stars we see in the night sky belong to the Milky Way. This is the star system, or galaxy, to which our Sun belongs. Like the Sun, the stars are great globes of searing, hot gas, which pour out energy as light, heat, and other forms of radiation. They are enormous distances away, and even in telescopes appear only as tiny points of light.

The light from a star is very feeble by the time it arrives at the Earth. Yet it can be made to reveal many of the star's secrets. Starlight can be gathered by a telescope and fed to an instrument called a spectroscope. The spectroscope splits the light into a spectrum in which a number of dark lines appear.

An enormous amount of information can be gained from the spectrum, such as the chemical composition of the star. The star's temperature can be found by observing which colors in the spectrum are brightest. The coolest stars are red, the hottest blue-white. The blue-white giant star Spica has a surface temperature of about 25,000°C (45,000°F), which is over four times the Sun's surface temperature. Stars of similar temperature display similar spectra, in which certain lines are prominent.

▶ Stars of all sizes, colors, and ages crowd into this photograph of part of the constellation of Sagittarius.

▼ This is a picture taken by the Hubble Space Telescope of one of the biggest stars we can see in the night sky. It is a red supergiant called Betelgeuse, in the constellation Orion.

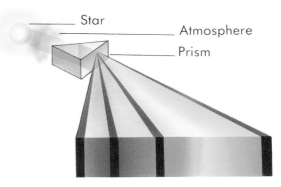

Stellar spectrum

Star — Atmosphere — Prism

Dark-line spectrum

The hot surface of a star gives out light of all wavelengths. If this light were passed through a prism, a complete spectrum, or rainbow, of color would be produced. But before the star's light reaches us, it has to pass through the star's outer atmosphere of cool gases. The various chemical elements in the atmosphere absorb certain wavelengths (or colors) from the light. As a result, sets of dark lines appear in the star's spectrum. From the positions of the lines, we can tell what elements caused them to appear. The lines in the Sun's spectrum are known as Fraunhofer lines for the German optician Josef von Fraunhofer, who first studied them.

Brightness, size, and speed

True brightness

We describe how bright a star is in the sky by its apparent magnitude. This is called apparent because it is a measure of the star's brightness as the star appears to us. It depends on how far away the star is as well as on the star's true brightness. For this reason, a close dim star may look brighter than a distant bright star.

To compare the true brightness of stars, we would have to look at all of them from the same distance. This is the basis of the scale of true brightness, or absolute magnitude. The absolute magnitude of a star is the apparent magnitude it would have at a distance of 10 parsecs, or about 33 light-years.

Sirius, the brightest star to our eyes, has an apparent magnitude of -1.45. But as stars go it is really not particularly bright, having an absolute magnitude of only +1.41. Deneb, the brightest star in Cygnus, has an apparent magnitude of +1.25, but an absolute magnitude of -7.3. It is one of the brightest of all the stars.

On the absolute scale the Sun is fairly dim, with a magnitude of only 4.8.

We can calculate a star's absolute magnitude from its apparent magnitude when we know how far away it is. This is because we know how brightness changes with distance.

Star size

The absolute magnitude is a measure of a star's luminosity, or how luminous it is. This depends on how hot the surface is and on the surface area. If we know the temperature of a star and its luminosity (absolute magnitude), we can work out its surface area and diameter.

We find that stars vary enormously in size. The Sun has a diameter of some 1,400,000 km (864,000 mi.). As stars go, it is fairly small, and is classed as a dwarf. Many stars have a similar size. But some are very much larger, and others very much smaller. Supergiant stars can be 500 times larger in diameter, whereas white dwarfs can be 100 times smaller.

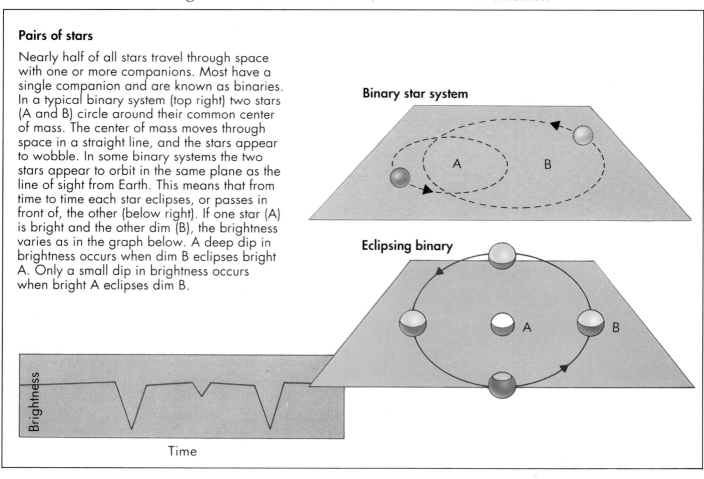

Pairs of stars

Nearly half of all stars travel through space with one or more companions. Most have a single companion and are known as binaries. In a typical binary system (top right) two stars (A and B) circle around their common center of mass. The center of mass moves through space in a straight line, and the stars appear to wobble. In some binary systems the two stars appear to orbit in the same plane as the line of sight from Earth. This means that from time to time each star eclipses, or passes in front of, the other (below right). If one star (A) is bright and the other dim (B), the brightness varies as in the graph below. A deep dip in brightness occurs when dim B eclipses bright A. Only a small dip in brightness occurs when bright A eclipses dim B.

Binary star system

Eclipsing binary

Brightness

Time

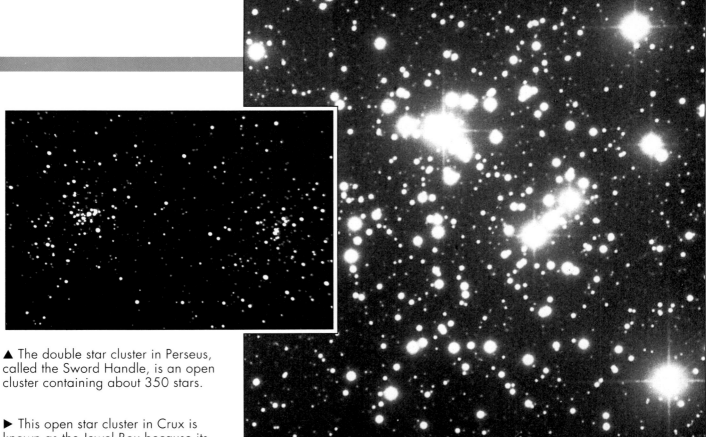

▲ The double star cluster in Perseus, called the Sword Handle, is an open cluster containing about 350 stars.

▶ This open star cluster in Crux is known as the Jewel Box because its stars flash different colors and sparkle like jewels.

Finding the mass of a star is difficult. But when a star is one of a pair in a binary system its mass can be found. It can be calculated from the distance the two stars are apart and the time it takes them to revolve around each other. Many stars have a mass similar to the Sun's. Some have only one-tenth the Sun's mass, and others are 10 times more massive.

Stars on the move

The stars appear to be fixed in the sky, but they are traveling rapidly through space. But most are so far away that their movement cannot be detected, even after hundreds of years.

In general, stars move toward us or away from us at an angle. So in time they should show a movement across our line of sight. We can detect this movement, called proper motion, for a few nearby stars, but it is only slight.

We can detect a star's movement toward or away from us by examining its spectrum. We call this movement radial motion. When a star is moving toward us, the lines in its spectrum are displaced toward the blue end. When a star is moving away, the lines are displaced toward the red end. The amount of blue or red shift is a measure of how fast the star is moving.

Globular clusters

Some of the most spectacular of all heavenly bodies are globular clusters. They consist of hundreds of thousands, even millions, of stars packed closely together in space in the shape of a globe. The brightest in the Northern Hemisphere is M13 (above), in Hercules. It is made up of about half a million stars. As in most globular clusters, the stars are very old. M13 is one of about 200 globular clusters found in the great halo that surrounds the center of our Galaxy.

The H-R diagram

As we have seen, two important features of a star are its temperature and its true brightness. Temperature is represented by the spectral class, and brightness by the absolute magnitude, or luminosity.

During the early part of this century, a pair of astronomers were investigating how the two are related. They were the Dane Ejnar Hertzsprung and the American Henry Russell.

In 1914 they published diagrams in which they plotted luminosity against spectral class for a number of stars. This type of diagram helped astronomers gain a new insight into the way stars are formed and how they gradually change during their lifetimes.

The illustration shows the Hertzsprung-Russell (H-R) diagram for some of the brightest and some of the dimmest stars. Diagonal lines across the diagram indicate the size of the stars compared with the Sun.

The most striking feature of the H-R diagram is that most stars lie along a diagonal band. We call this band the main sequence. Stars on the main sequence shine steadily. They include, about halfway down, our own Sun. Stars at the top of the band are very hot, with a surface temperature around 52,000°C (about 93,632°F). They are also very bright – up to 1.4 million times brighter than the Sun. In contrast, the dim red stars at the bottom of the band are only about one ten-thousandth as bright.

Giants and dwarfs

A number of stars lie off the main sequence. In the upper right are the cool but highly luminous group known as red giants because of their enormous size. The stars above them are even bigger and brighter supergiants. In contrast, below the main sequence are a group of hot, but tiny, white dwarf stars. These represent a late stage in the life of stars similar to the Sun.

▶ On the Hertzsprung-Russell diagram, the stars are bigger and brighter toward the top of the diagram. Going from right to left they get hotter. The inset diagrams show the very great differences in the sizes of stars. A supergiant has a diameter several hundred times that of the Sun. On the other hand the Sun is about 100 times bigger than a white dwarf. A neutron star is a thousand times smaller still, only about 20 km (roughly 10 mi.) across. A black hole may be even smaller.

18

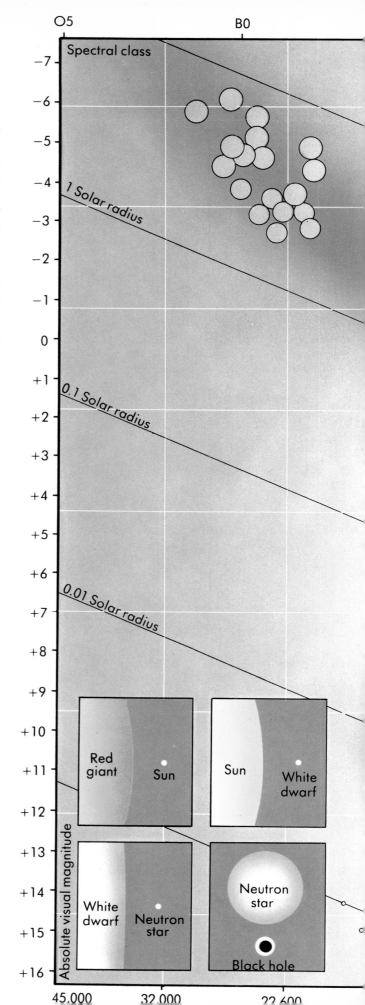

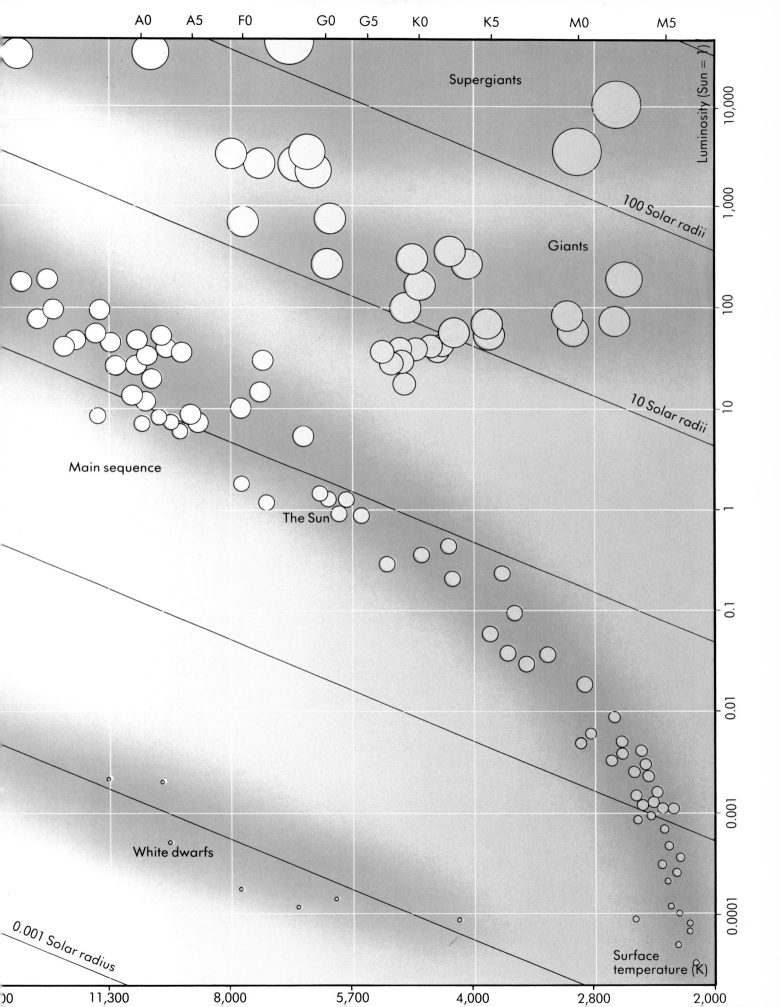

Variable stars

Most stars shine steadily. But some vary in brightness from time to time. We call them variable stars.

One of the first variables discovered was Mira, in the constellation Cetus. It changes in brightness between about magnitude 3 and magnitude 10 over a period of about 331 days. It is classed as a long-period variable. Many other Mira-type stars are now known. They are all red giants.

Many other variable stars brighten and dim less noticeably than Mira and in a much shorter period. But they do so with absolute precision. The first to be discovered was Delta Cephei, in the constellation Cepheus. It gave its name to the class of short-period variables which are known as Cepheids.

Typical, or classical, Cepheids have a period of from 1 to 60 days. They are young, massive giant arid supergiant stars. Their period is directly related to their magnitude.

In contrast, there are many other groups of variables that vary in their brightness quite irregularly. The supergiant star Betelgeuse is an example. It changes in brightness over a period of roughly five years. The so-called eruptive variables, such as UV Ceti, change in brightness once or twice a day.

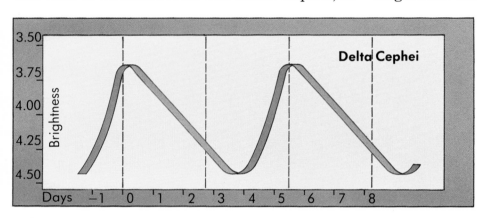

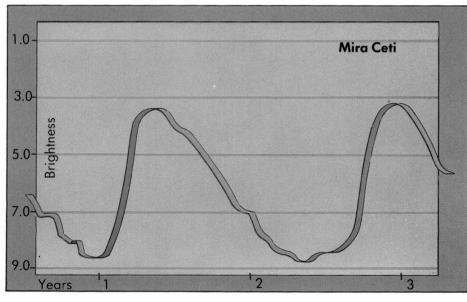

▲ The brightness of Cepheid stars varies in a regular way. Delta Cephei itself (top graph) varies in brightness between magnitudes 3.5 and 4.3 in a period of precisely 5.37 days.

▲ The variation in brightness of Mira Ceti (above) is irregular over a much longer period. It varies between about magnitudes 3 and 10 about every 11 months.

Noted stargazer

The American astronomer Henrietta Leavitt became famous for her work with Cepheid stars at Harvard Observatory. In 1912 she made the discovery of the period-luminosity law for these stars. She found that the period of a Cepheid is directly related to its absolute brightness, or luminosity. The longer its period, the greater is its luminosity. This law enables us to measure the distances to Cepheids.

Bright nebulae

The space between the stars is not completely empty. It contains minute traces of gases and tiny grains of dust. In places this interstellar matter clumps together to form denser clouds, which we call nebulae.

Some nebulae contain very hot stars and shine brilliantly. Energy absorbed from the stars causes the gases present to emit light. We call this type emission nebulae. We can see one with the naked eye in the constellation Orion. It is called the Great Nebula in Orion, or M42.

When a star lies outside a gas and dust cloud, the cloud may reflect starlight. We then see it as a reflection nebula.

▼ A photograph of part of a gigantic gas and dust cloud that lies approximately 700 light-years away near the star Rho Ophiuchi (top left). The dark regions show where the cloud is thickest. The blue area is a reflection nebula, and the pink color is due to the presence of the gas hydrogen.

▼ When a star lies in front of a cloud of dust, the dust reflects light and we see a bright reflection nebula (left side of picture). If the cloud lies between us and the star, the star's light is blocked and we see a dark nebula (right side of picture).

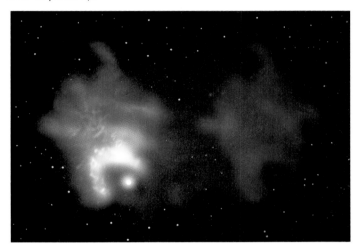

Between the stars

Dark nebulae

There are many dust clouds in the heavens that are not lit up by nearby stars. We can see them as dark patches against a starry background. They blot out the light of the stars behind them. We call them dark nebulae. Two of the best-known ones, the Coalsack and the Horsehead, are pictured on these pages.

Some dark nebulae, called globules, are small, round, and much denser than ordinary nebulae. Astronomers believe that they will one day turn into stars.

A dusty disk

Dust does not only occur clumped together in clouds. It is also scattered haphazardly in space between the stars. It has the effect of dimming their light. There is much dust in the disk of our Galaxy, the Milky Way. It masks our view of the center of the Galaxy.

Interstellar dust appears to be made up of specks of carbon or of silicates (like many Earth rocks). Often the specks are coated with ice. Traces of other substances are also found in between the stars, including water, alcohol, and glycine. Glycine is an amino acid, one of the building blocks of life.

▼ The constellation of the Southern Cross stands out in this view of part of the Milky Way in the Southern Hemisphere. Near the two brightest stars of the Cross is what appears to be a dark "hole." In fact it is a dark nebula, called the Coalsack.

Molecules between the stars

Astronomers have discovered more than 50 different chemicals in the gas and dust clouds between the stars. Some of their molecules show up in the spectra of the light the clouds give off. Others make themselves known by the radio waves they emit. It is interesting that there are several carbon compounds among the molecules, because such compounds are the basis of life on Earth.

Name	Formula
Cyanogen	CN
Hydroxyl	OH
Ammonia	NH_3
Water	H_2O
Formaldehyde	H_2CO
Carbon monoxide	CO
Hydrogen	H_2
Hydrogen cyanide	HCN
Methanol (methyl alcohol)	CH_3OH
Methanoic acid (formic acid)	HCO_2H
Silicon monoxide	SiO
Acetaldehyde	CH_3CHO
Hydrogen sulfide	H_2S
Dimethyl ether	CH_3OCH_3
Ethanol (ethyl alcohol)	CH_3CH_2OH
Sulfur dioxide	SO_2
Ethyl cyanide	CH_3CH_2CN
Nitric oxide	NO

▶ Of all the dark nebulae, none is better known or better named than the Horsehead, in Orion. The dark cloud of dust in the shape of a horse's head stands out vividly against a bright emission nebula.

Birth and death of stars

Stars are born in the great clouds of gas and dust that are scattered throughout the galaxies. When their nuclear furnaces light up, they start to shine. Some glow feebly, while others blaze like great celestial beacons. The bigger and brighter a star is, the shorter its lifetime, and the more spectacular its death. Stars like the Sun meet quite a peaceful end, shrinking into a superdense white dwarf as small as our planet. But more massive stars condense to a size larger than the Earth. They blast themselves apart in a mighty supernova explosion and briefly shine nearly as bright as a galaxy.

SPOT FACTS

• In about 5 billion years' time the Sun will swell up into a red giant. It will certainly expand beyond the orbit of Mercury and possibly even beyond that of Venus.

• The first white dwarf to be recognized (in 1915) was the faint companion of the Dog Star, Sirius, brightest star in the heavens. Properly termed Sirius B, it is often called the Pup. It is about the size of Earth, but it is 350,000 times more massive.

• In 1982 astronomers discovered the first millisecond pulsar PSR 1937 + 21. It flashes on and off every 1.56 milliseconds (thousandths of a second) as it spins 642 times per second.

• The Hubble Space Telescope has found over 100 newly born stars in the Orion Nebula, each still surrounded by a disk of material that may eventually form planets. They are called "proplyds" for "proto-planetary disks."

• Within a star, the chances against one hydrogen atom fusing with another to create energy are billions to one. Yet there are so many hydrogen atoms in a star that they fuse together millions of times a second.

• Red supergiant stars such as Betelgeuse can be up to 1,000 times the diameter of the Sun.

• The first brown dwarf star to be detected was observed by the Hubble Space Telescope in 1995. Called Gleise 229B, it is about 50 times the mass of the planet Jupiter.

• Brown dwarfs do not shine by nuclear reactions, but glow dull red through the force of gravity, which causes them to shrink.

• Red giant stars are the major source of carbon and water in the Universe. These are given off from their surfaces, and collect in dust clouds. The dust particles probably have a carbon core surrounded by ice.

• The only place in our Galaxy hot enough for elements heavier than iron to form is in a supernova explosion. Elements such as gold, silver, and platinum were all created in supernovas.

• A supernova explosion distributes heavy elements throughout space and can create shock waves in dust clouds, triggering off the formation of new stars.

• Black holes do not suck everything nearby into them; they have exactly the same effect on nearby material as an ordinary star of the same mass.

• If anyone were to venture very close to a black hole, the pull of its gravity on one end of their body would be much greater than that on the other. So they would be pulled apart. This is known as "spaghettification."

• If a supernova explosion were to take place close to Earth, the energetic particles given off would wipe out most life. Fortunately there are no suitable stars nearby.

• The material from the explosion of supernova 1987A was ejected from it at 17,000 km (10,563 mi.) per second.

A star is born

Nobody knows quite what triggers a cloud of gas and dust to turn into a star. As the cloud collapses, energy is released, which causes it to heat up. The center of the cloud reaches 10 million degrees Celsius or more.

Most of the gas in interstellar clouds is hydrogen. And at such high temperatures, the hydrogen atoms start to combine, or fuse together. This fusion reaction produces enormous amounts of energy as light, heat, and other radiation. When this happens, the collapsing cloud starts to shine as a star.

The outward "pressure" of the radiation coming from the core of the new star acts against the matter that is collapsing under gravity. Eventually the two balance each other, and the collapse ceases. The star settles down and begins to shine steadily. It takes a star the size of the Sun about 50 million years to reach this state.

A Sun-sized star shines steadily for about 10 billion years, until the hydrogen fuel in its core is used up. The star then begins to collapse again under gravity. The heat triggers hydrogen fusion in the gassy shell surrounding the core. The shell heats up, causing the star to expand and brighten. But the core continues to shrink and get hotter.

The Eagle Nebula star nursery

A picture from the Hubble Space Telescope of part of a giant hydrogen cloud known as the Eagle Nebula, in the constellation of Sepens, 7,000 light-years away.

Within this dense cloud are newborn stars that emerge from the fingers of gas at the top of the cloud when the gas is boiled away by the light from other stars.

Stellar life cycles

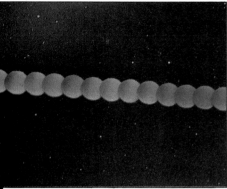

◄ The way a star lives and dies depends on how massive it is. A brown dwarf (3) never shines brightly. A star such as the Sun spends much of its life on the main sequence (5). The Sun's main-sequence lifetime is about 10 billion years. More massive stars (8, 12) have a much briefer life, and blast themselves apart spectacularly as a supernova (10).

▼ A planetary nebula called the Cat's Eye nebula, taken by the Hubble Space Telescope. Near the end of its red-giant stage, a star may blow some of its matter out into space. Through a telescope the matter looks much like a planet and so early astronomers called it a planetary nebula.

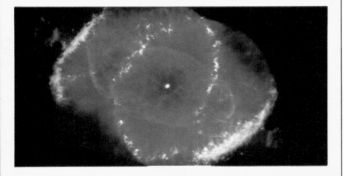

When the temperature in the shrinking core of a star reaches 100 million degrees Celsius, another fusion reaction is able to take place. This reaction changes the nuclei of helium atoms into carbon nuclei. It is known as the triple-alpha reaction, because it combines three helium nuclei, otherwise known as alpha particles. The triple-alpha fusion process provides the energy to keep the expanded star shining, as a red giant.

A Sun-sized star expands to up to 100 times its initial diameter while becoming a red giant. But in time all the helium in the core is used up, and the star again begins to shrink. Eventually gravity crushes the matter in the star into a small planet-sized body of immense density. This body is called a white dwarf.

Stars with a smaller mass than the Sun have a longer life. Those with a larger mass have a shorter life. Some massive stars may live for only a few million years. They burn up their fuel rapidly, then swell up into a supergiant. Finally they become a brilliant supernova and blast themselves apart.

Gas from stars

Red giant

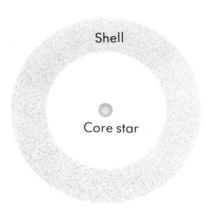

Shell

Core star

▲► An old red giant star (above right) often blows out gas, which forms an enveloping shell around it. The shell expands as time goes by. It absorbs energy from the central star and glows as a planetary nebula.

► The Ring nebula is one of the most beautiful of the ring-shaped planetary nebulae. Looking like a colorful smoke ring, it lies in the constellation Lyra, and measures about 0.6 light-years in diameter.

Violent death

Stars with several times the mass of the Sun meet their end in the most spectacular way. They grow to an enormous size as a supergiant and then blast themselves apart, exploding as supernovae.

When a star becomes a supernova, its brightness increases many millions of times. For a while it may even shine more brightly than a whole galaxy. In our Galaxy three supernovae have been seen during the last 1,000 years. Chinese astronomers spotted one in AD 1054, Tycho Brahe studied one in 1572, and Johannes Kepler saw one in 1604. Supernovae are so brilliant that we can see them in other galaxies.

What happens to a star after it has exploded as a supernova depends on how massive it is. One with a mass of more than seven times the Sun's mass becomes a neutron star. Stars more massive still become black holes.

A neutron star is formed when the central core of an exploding star collapses under gravity. As it gets smaller and smaller, the protons and electrons in its atoms are crushed together and form neutrons. The star turns into a "sea" of neutrons that is incredibly dense: a teaspoonful would weigh 100 million metric tons! Astronomers think that the bodies known as pulsars are neutron stars that rotate rapidly. As they do so, they produce a beam of radio waves. This sweeps round in space like the beam from a lighthouse. We receive radio pulses when the beam flashes in our direction. One of the first pulsars to be discovered was the Crab pulsar, in the Crab nebula.

Brighter than a billion Suns

Near the end of their life, heavy stars blow themselves apart as a supernova. They may first blaze more brightly than a billion Suns. On February 23, 1987, a star erupted into a supernova in the Large Magellanic Cloud (left), the galaxy nearest to Earth. It became the brightest supernova seen for nearly 400 years (below left).

A supernova blasts vast amounts of gas and dust into the surrounding space, which forms an expanding cloud. The gas cloud we call the Crab nebula (below) is what remains of a supernova first seen in the year 1054.

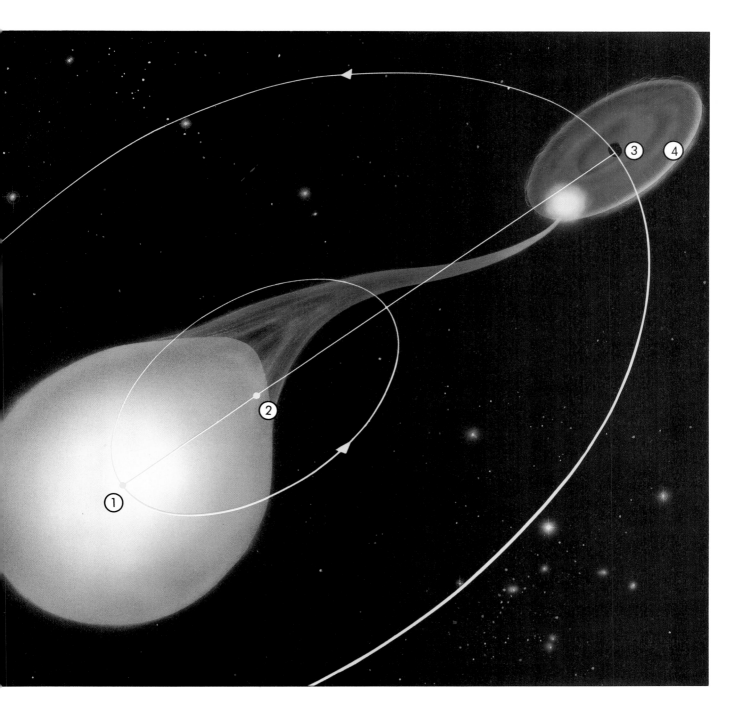

Very heavy stars do not stop collapsing even when they shrink to the neutron-star stage. Their gravity is so great that the collapse continues. The matter which they contain is eventually crushed into a point, known as a singularity. In the region around this point gravity is so intense that nothing, not even light, is able to escape from it. That is why astronomers call such a region a black hole.

▲ Astronomers believe that there may be a black hole in a binary star system in the constellation Cygnus. They think that the system is made up of a supergiant star (1) and a black hole (3), rotating around the center of mass (2). The star system is known as Cygnus X–1 because of the powerful X-rays it gives out. In such a system, X-rays are produced as matter, torn from the supergiant. They spin into a spiral disk of matter (4), racing furiously around the black hole.

Galaxies galore

Stars are not spread about haphazardly in space. They gather together into great spinning star islands, or galaxies. All the stars we see in the sky belong to our home galaxy, which we call the Milky Way, or just the Galaxy. It is one of perhaps 100 billion galaxies in the Universe, many containing billions of stars. Most galaxies lead relatively peaceful lives, producing a steady output of light. Some, however, are noticeably more active. They pour out up to a million times more energy than normal, particularly as radio waves.

SPOT FACTS

• The halo that surrounds our Galaxy may itself be surrounded by a "dark halo" of invisible matter. It could extend more than 200,000 light years from the Galaxy's center.

• The Sun orbits around the center of the Galaxy at a speed of about 900,000 km per hour (nearly 600,000 mph). It is now traveling in its 23rd orbit.

• A belt of fast-moving gas connects the Galaxy with its two nearest neighbors in space, the Large and Small Magellanic Clouds.

• The Andromeda galaxy is the most distant object we can see in the heavens with the naked eye. Its light takes 2.2 million years to reach the Earth. The light that we see now began its journey when our earliest ancestors were living on Earth.

• Only about a third of all nearby galaxies are spirals like our own and the Andromeda galaxy. Over half are irregular in shape.

• Most of the stars we see in the sky with the naked eye are within a few hundred light years of the Sun.

• The center of our Galaxy is hidden from view by dust clouds. Infrared and radio observations suggest that it contains a black hole.

• The nucleus of the Galaxy consists of tightly packed stars with a combined brightness of a billion times that of the Sun.

• The supermassive black holes believed to be at the centers of quasars and active galaxies probably have masses around a billion times that of the Sun.

• The motions of the galaxies in the Local Group – the name given to the cluster of galaxies that includes our own Galaxy – suggest that there is far more material in it than is visible. Other galaxy clusters also seem to contain "missing mass." Astronomers are still debating what form this mass may take.

• A mysterious concentration of mass known as the "Great Attractor" has been found at a distance of 300 million light years from our Galaxy. It consists of a tight group of over 600 galaxies that are pulling nearby galaxies toward them.

• A band of galaxies stretching for 500 million light years across the sky is known as the "Great Wall."

• As well as clumping together in clusters, there are regions of space known as voids that seem to contain few galaxies. The reason for these phenomena is still being investigated.

• Gravitational lenses are being used as an important means of probing the Universe. Some lenses split the light from distant quasars into a distinctive cross shape, known as an "Einstein Cross," or into a delicate arc of light.

• Many astronomers believe that quasars are the most distant objects that can be observed. But a minority believe they are comparatively small, close objects, which give the impression of being distant.

• The jets from some quasars appear to be moving many times faster than the speed of light, which is contrary to Einstein's Theory of Special Relativity. But the effect is now known to be an optical illusion when the material is traveling toward us at close to the speed of light.

The Milky Way

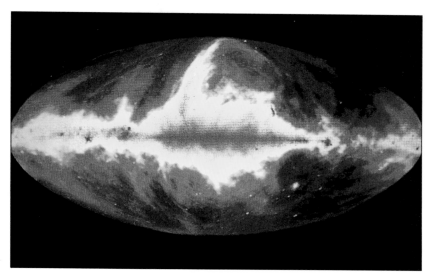

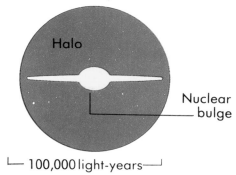

Dimensions of the Galaxy

Halo

Nuclear bulge

— 100,000 light-years —

◀ This map of our Galaxy was prepared using radio waves rather than light waves. Red shows regions with strong signals.

On a moonless night, you can see a fuzzy band of light arcing across the heavens. We call it the Milky Way. In the Northern Hemisphere it passes through the easily recognized constellations Cassiopeia and Cygnus. In the Southern Hemisphere it passes through the unmistakable Scorpius and Crux, the Southern Cross. The brightest part lies in Sagittarius.

When viewed through a telescope, the Milky Way turns out to be a region containing millions of faint stars seemingly packed close together. This is because when we look at the Milky Way, we are seeing a cross section of our own Galaxy. The stars are really far apart. They just appear to be close together because of the way we view them from the Earth.

The Galaxy takes the form of a flattish disk with a bulge (nucleus) in the middle. The 100 billion stars it contains are spread out on the disk. In practice, the stars in the disk group together on arms that spiral out from the center. The whole Galaxy rotates around the center, but not at a uniform speed. Stars at the center travel faster than those farther out. The Sun (30,000 light-years from the center) takes 225 million Earth-years to make one rotation. This period of time is called a cosmic year.

▶ If we could view our Galaxy from far out in space, it would look something like this, the famous Andromeda galaxy. The Andromeda galaxy is one of the few we can see with the naked eye. It is bigger than our Galaxy, but has a similar spiral shape.

Spirals, ellipticals, and irregulars

Our Galaxy, the Milky Way, is one of many millions of spiral galaxies in the Universe. The others are of much the same shape and contain much the same mix of stars, clusters, gas, and dust. Some are smaller than the Milky Way; others are larger. One of our galactic neighbors, the Andromeda galaxy, contains more stars than our Galaxy.

Close to Andromeda in the sky are two much smaller galaxies. They are elliptical in shape, and have no spiral arms. Ellipticals are made up mainly of old stars, unlike spirals, which contain many young stars.

The galaxies nearest the Earth, however, have no distinct shape, and are classed as irregulars. They are called the Large and Small Magellanic Clouds (LMC and SMC). They can both be seen with the naked eye in far southern skies. The LMC measures 30,000 light-years across, a third of the size of our Galaxy. It appears to be twice as big as the SMC.

Classifier of galaxies

The American astronomer Edwin Hubble is pictured here at the controls of the 100-inch reflector at Mount Wilson Observatory in California. (The telescope's mirror is 2.54 m, or 100 in., in diameter.) In 1923 he began devising a method of classifying the galaxies. His system grouped galaxies according to their shape: as ellipticals, spirals, and barred spirals.

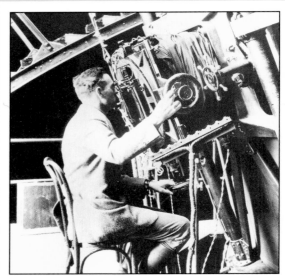

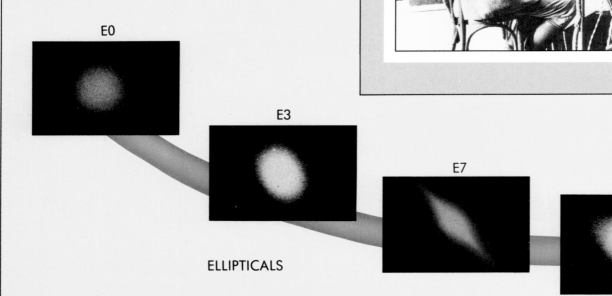

E0

E3

E7

S0

ELLIPTICALS

Portrait of the galaxies

The pictures show examples of the types of galaxies Hubble included in his historic classification. He thought that the galaxies evolved, in the sequence shown, from ellipticals into spirals. But astronomers no longer believe this. They are not certain how the different types of galaxies are related.

Elliptical galaxies (E) are described by a number from 0 to 7, which indicates how flattened they are.

Spiral galaxies (S) have a central bulge, or nucleus, from which a number of arms curve out. They are classed as a, b, or c, depending on how open the arms happen to be. So galaxies are similar to spirals, but have no arms.

Barred-spiral galaxies (SB), on the other hand, have spiral arms that come out of the ends of a line of stars (bar) through the nucleus. They are classed as a, b, or c.

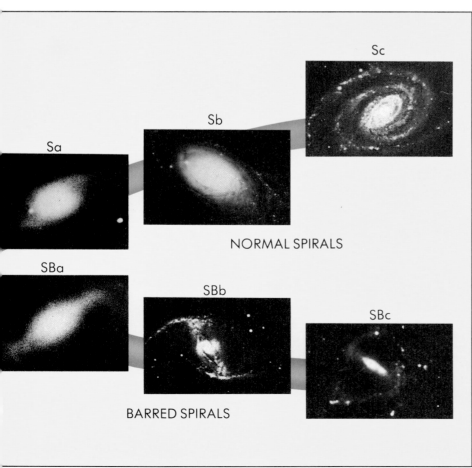

Sc

Sb

Sa

NORMAL SPIRALS

SBa

SBb

SBc

BARRED SPIRALS

▲ The galaxy M33 is number 33 in a catalog of "nebulae" drawn up by the French astronomer Charles Messier in 1774. It is a spiral galaxy, class Sc, which has wide-open arms that spiral from the nucleus. The stars in the spiral arms (blue) are young. The nucleus is choked with dust (orange), lit by old yellow stars.

▼ The Small Magellanic Cloud is our second-nearest neighboring galaxy. It is a milky white patch, visible only in the Southern Hemisphere.

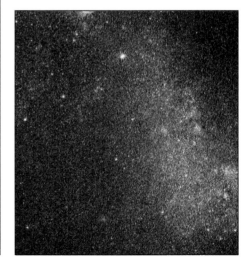

Active galaxies

There are some galaxies that do not fit neatly into any class. Many are spiral galaxies with unusual features. Some may be colliding or have collided at some time in the past. For example, astronomers think that the Cartwheel galaxy was a spiral whose center was knocked out in a collision hundreds of millions of years ago.

Seyfert galaxies

Another type of spiral galaxy is notable for having a very bright nucleus and very faint spiral arms. Such galaxies were first studied in the 1940s by the American astronomer Carl Seyfert, and are now termed Seyfert galaxies. Because their centers shine so brightly, they can sometimes be mistaken for stars.

Seyfert galaxies are one type of active galaxy, one that has an unusually powerful energy source at its center. Other active galaxies are very strong emitters of radio waves, and are termed radio galaxies.

The study of radio galaxies started when astronomers began "tuning into" the heavens with radio telescopes. One of the first radio sources found was located in the southern constellation Centaurus and named Centaurus A. In 1949 Australian astronomers identified this source with the bright galaxy NGC 5128.

Twin lobes

In photographs made with light waves Centaurus A looks like an ordinary elliptical galaxy, as do most radio galaxies. But at radio wavelengths they can outshine an ordinary galaxy by up to a million times. The radio waves do not come from the center of the galaxy. They come from twin lobes, regions of space on each side of the galaxy. In the case of Centaurus A, the lobes extend over 2,500,000 light-years. Yet the galaxy is only 30,000 light-years across!

Images of the radio galaxy M87 indicate how the radio waves are emitted. They show a jet, or stream, of matter shooting out of the nucleus. The jet is a high-speed beam of electrons. As they pass through the magnetic field surrounding the galaxy, they are forced into a spiral path around the magnetic lines of force. This constant change of direction triggers the radiation we detect as radio waves.

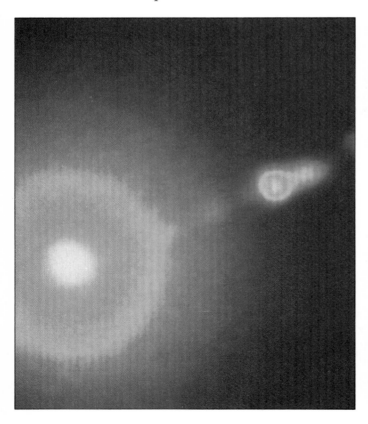

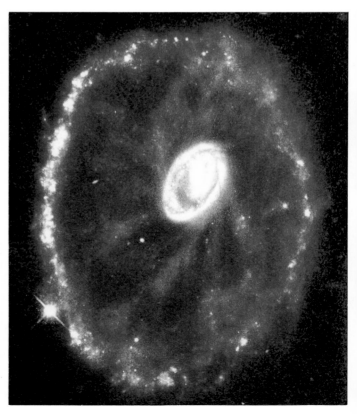

▲ ▶ The galaxy Centaurus A is crossed by a broad dust lane. In most of the elliptical central regions, there are mainly old yellow stars, with younger blue ones around the edges of the lane. At right is a radio map of the same galaxy. The image reproduced above would fit within the pink circle in the middle.

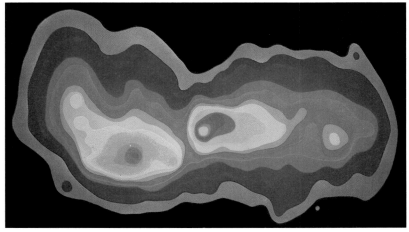

◄ (Far left) M87, one of the nearest active galaxies, has a long "jet" extending from it. The picture was made using a radio telescope. Astronomers believe there is a huge black hole at the galaxy's center. (Left) The Cartwheel Galaxy, photographed by the Hubble Space Telescope, is the result of a collision between galaxies.

Quasars

The most intriguing of all the active galaxy-type objects in the Universe are quasars. The first two to be discovered were not given names, but coded as 3C 48 (in 1960) and 3C 273 (in 1962). They were strong radio sources whose positions in the sky matched those of two faint blue stars. The spectra of the stars, however, were quite unlike any seen before.

Astronomers now know the reason for this. The lines in the spectrum of these starlike objects are shifted to the red by an enormous amount. In other words the objects must be very far away. 3C 273 proves to be more than 2 billion light-years away. No star can be seen from this distance. Therefore it cannot be an ordinary star, even though it looks like one. And for it to be visible from such a distance, it must be hundreds of times brighter than an ordinary galaxy. Astronomers have since discovered more than 7,500 other quasars, or quasi-stellar radio sources.

Quasars do not shine steadily like ordinary galaxies. They vary in brightness over periods of days or years. For this reason they cannot possibly be as big as an ordinary galaxy. For if a quasar changes in brightness in a year, say, it cannot be more than one light-year across. And if it changes in brightness in a day, it cannot be more than a light-day across. From its variation in brightness, 3C 273 works out to be less than one-hundredth of a light-year across, which makes if only one ten-millionth the size of a typical galaxy!

However, it seems that quasars are not separate bodies. They appear to be eruptions at the center of massive galaxies. The rest of the galaxies are too faint to be visible at the distances involved.

The powerhouse

What kind of energy source could make quasars the size of the Solar System put out the power of hundreds of galaxies? There seems only one possibility – a massive black hole.

A black hole is created when an aging star collapses. It is a region of space with superhigh gravity, which swallows matter like a cosmic vacuum cleaner. It is surrounded by a rapidly rotating disk of hot gas. Matter attracted by the hole's enormous gravity acquires great amounts of energy. This is released as radiation when the matter plows into the disk.

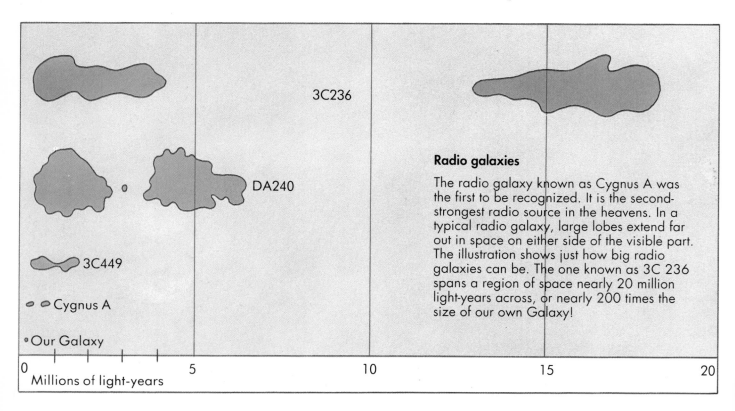

3C236

DA240

3C449

Cygnus A

Our Galaxy

0   5   10   15   20
Millions of light-years

Radio galaxies

The radio galaxy known as Cygnus A was the first to be recognized. It is the second-strongest radio source in the heavens. In a typical radio galaxy, large lobes extend far out in space on either side of the visible part. The illustration shows just how big radio galaxies can be. The one known as 3C 236 spans a region of space nearly 20 million light-years across, or nearly 200 times the size of our own Galaxy!

36

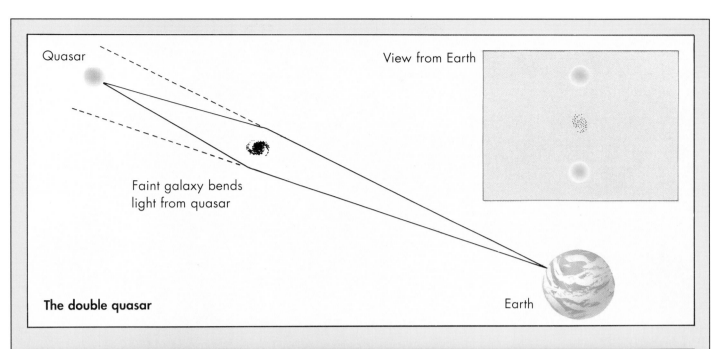

Quasar

View from Earth

Faint galaxy bends
light from quasar

The double quasar

Earth

Seen from the Earth a quasar can appear as a multiple
quasar when it lies behind another galaxy, which acts as
a "gravitational lens." The diagram above shows a
double quasar; this photograph shows an "Einstein's
Cross," when a quasar's image is split into four.

Clusters of galaxies

Just as stars cluster together in space in galaxies, so the galaxies themselves tend to cluster together. Our own Galaxy, the Milky Way, has two close companion galaxies, the Large and Small Magellanic Clouds. All three galaxies form part of a cluster of about 30 galaxies known as the Local Group.

This group contains three large spirals: the Milky Way, M33, and the Andromeda galaxy. M33 lies about 2.4 million light-years away, about 200,000 light-years farther than Andromeda. The Magellanic Clouds are two of four small irregular galaxies in the group. Most of the galaxies, however, are ellipticals and they are smaller still.

Two of the small elliptical galaxies are surrounded by clusters of millions of stars, known as globular clusters. The small ellipticals are found in the southern constellations of Sculptor and Fornax.

The only large elliptical in the Local Group, Maffei I, is probably as massive as our own Galaxy. It lies 3.3 million light-years away. The Local Group is quite a small cluster of galaxies. The nearest major cluster contains between 1,000 and 2,000 galaxies. It is located in the constellation Virgo and is centered on the powerful active galaxy M87.

But most clusters contain only between about 100 and 400 galaxies. The grouping of galaxies also seems to occur on an even larger scale. The clusters apparently form part of massive superclusters. Our Local Group forms part of a supercluster that is dominated by the huge Virgo cluster. It contains about 100 clusters in a region of space about 250 million light-years across. A cluster of galaxies in Hercules, some 600 million light-years away, is part of an even larger supercluster. It is so large that it spans a third of the sky.

▼ The Hubble Space Telescope has been used to record the light from one small part of the sky for many hours. The result shows the faintest galaxies ever detected. Almost every object on this picture is a distant galaxy, revealing a wide range of different types and sizes.

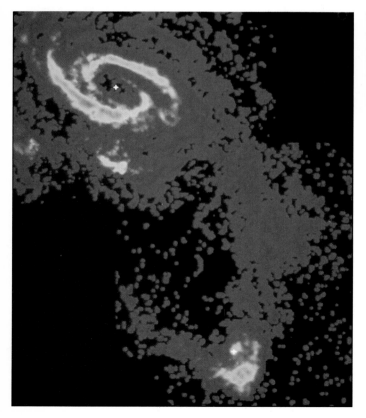

▲ Abell 1060 is a typical cluster of galaxies.

► Our Galaxy is one of the biggest members of the Local Group (numbers), which in turn forms part of the local supercluster (letters).

Local Group
1 Leo II
2 Sculptor system
3 Small Magellanic Cloud
4 Leo I
5 Draco system
6 Large Magellanic Cloud
7 Our Galaxy
8 NGC 6822
9 Ursa Minor system
10 NGC 147
11 NGC 185
12 IC 1613
13 M33
14 M31 (Andromeda)
15 M32

Local supercluster
A Virgo III cloud
B Virgo II cloud
C Crater cloud
D Virgo I cloud
E Leo cloud
F Canes Venatici cloud
G Canes Venatici spur

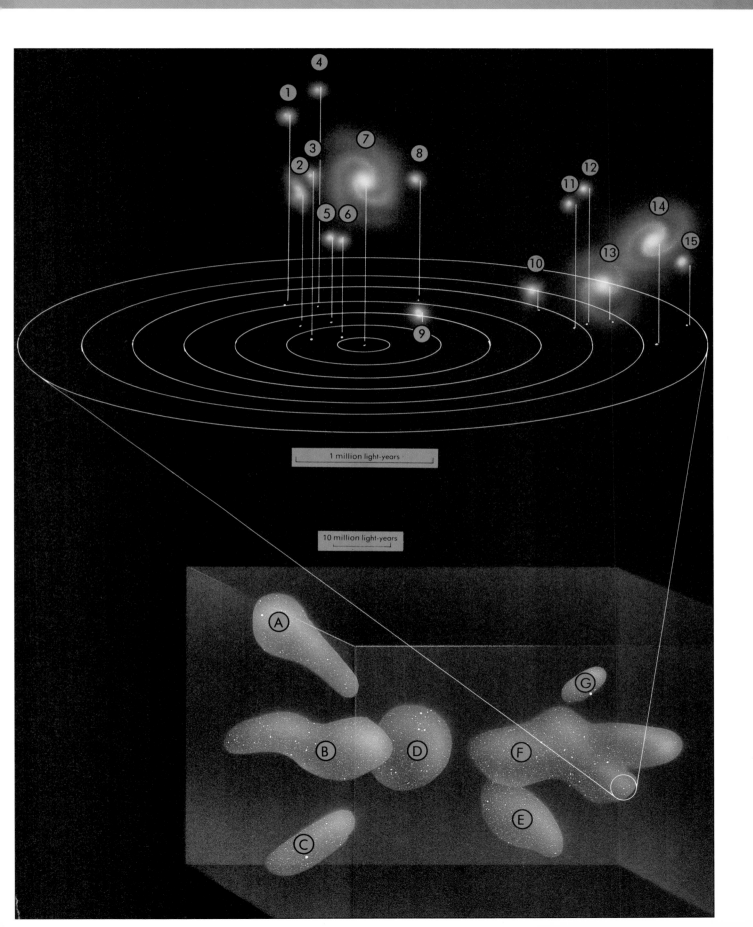

1 million light-years

10 million light-years

Big Bang, Big Crunch

How did the Universe begin, and when? How has it evolved? How will it end, and when? These are the kinds of questions astronomers have been trying to answer for centuries. Cosmologists, the astronomers who study such things, now think they have many of the answers. They see in the expansion of the Universe evidence that it began in a gigantic explosion. And they have worked out how it has evolved from the first millisecond of its existence to the present day. They are not so sure how the Universe will end, but the signs are that it will go on expanding forever until the stars and galaxies fade away.

SPOT FACTS

• Astronomers believe that the speeds at which galaxies are receding from the Earth increase by about 55 km (34 mi.) per second for each megaparsec (3,260,000 light years) of distance – a rate known as the "Hubble constant."

• Some recent measurements of the Hubble constant give it a value of around 80 km (50 mi.) per second rather than 55 km (34 mi.). But this would make the Universe younger than its oldest stars. This is one of the major puzzles of astronomy.

• The shift in a star's light toward the red end of the color spectrum indicates the speed at which the star is traveling away from us. From their red shifts, the most distant quasars appear to be traveling at over 270,000 km (169,000 mi.) per second, or more than 90 percent of the speed of light. This places them more than 13 billion light years away.

• Astronomers believe that the same laws of physics apply everywhere in the Universe. For example, the speed of light and the relative value of gravity are the same everywhere.

• Life originated on Earth almost as soon as its surface had cooled sufficiently.

• Both the age of the Universe and the number of stars is so great that there is a strong chance that life has arisen on billions of other planets.

• Despite attempts to listen for intelligent radio signals from planets of other stars, none has yet been found.

• When the Universe is 100 times older than it is now, all the stars will be dead and the galaxies will be fading.

• The first atomic nuclei began forming just three minutes after the Universe was born.

• The discovery in 1965 of the Universe's background radiation by Arno Penzias and Robert Wilson was made while they were trying to eliminate annoying interference in an antenna that belonged to the Bell Telephone Company.

• Slight variations have been found in the brightness of the background cosmic radiation. These variations are believed to result from ripples in the Big Bang, which eventually formed into clusters of galaxies.

• Research into the Big Bang, at the largest scale of the Universe, is closely linked to the work of particle physicists who investigate the tiniest properties of matter.

• Any slight change in the values of the mass or charge of particles would mean that the Universe could not exist in its present form, with quarks combining to make particles, which in turn form elements.

• Some scientists believe there may be other universes with such different properties that we can never observe them.

• If the Universe were not expanding, and if it were infinite in size, we would see a star or galaxy in every direction, and the sky would everywhere be the same brightness as the surface of the Sun.

An expanding Universe

In 1914 the American astronomer Vesto Slipher began studying the spectra of galaxies. He found that all of them, with the exception of the Local Group, had red shifts. The red shifts indicated that all the galaxies were moving away from us. At the time no one was sure if the galaxies were part of the Milky Way or not. In the 1920s, however, fellow American Edwin Hubble showed without doubt that the galaxies did lie outside it. So, with all the galaxies rushing away, it appeared that the whole Universe was expanding. All evidence since has confirmed that this is happening.

So it follows that in the past the Universe must have been smaller. Working backward, there must have been a time when all the matter in the Universe was packed together in one place. This is the reasoning behind the most widely accepted theory about how the Universe began. It is thought that the Universe came into being as a result of an explosion, or Big Bang, which set in motion the expansion that we observe. Astronomers have worked out that the Big Bang must have occurred about 15 billion years ago.

During his study of the spectra of galaxies, Hubble also discovered another interesting thing. The farther away galaxies are, the greater is their red shift and the faster they are moving. And he worked out a relationship between speed and distance, which became known as the Hubble constant.

Measuring the amount of red shift and applying the Hubble constant provides the only method of finding out the distance to the farthest galaxies and the remote quasars.

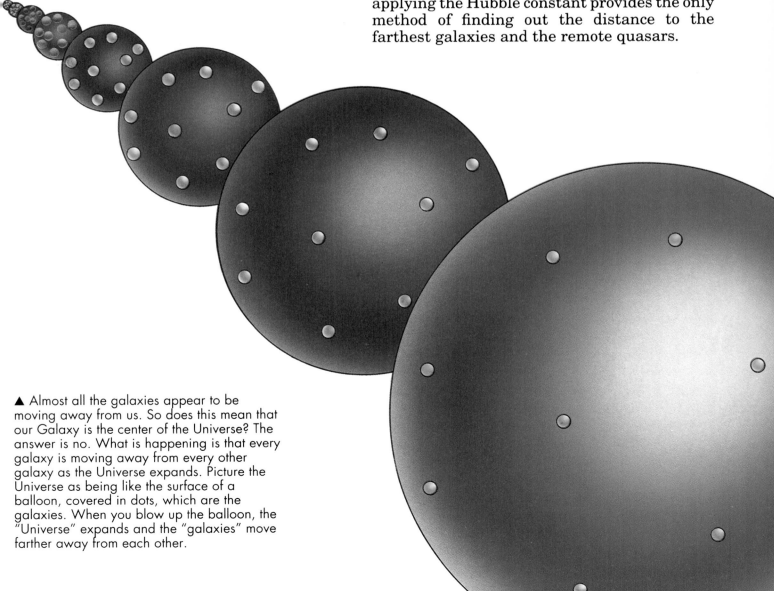

▲ Almost all the galaxies appear to be moving away from us. So does this mean that our Galaxy is the center of the Universe? The answer is no. What is happening is that every galaxy is moving away from every other galaxy as the Universe expands. Picture the Universe as being like the surface of a balloon, covered in dots, which are the galaxies. When you blow up the balloon, the "Universe" expands and the "galaxies" move farther away from each other.

After the Big Bang

We cannot think of the Big Bang as being like an ordinary explosion, which scatters material into the surrounding space. There was nothing before the Big Bang - no matter, no energy, no space, and no time. The Universe did not exist. Matter, energy, space, and time came into being with the Big Bang. The Universe was born then, some 15 billion years ago.

Astonishingly, astronomers think they know how the Universe has evolved from the very beginning. A millionth of a second after the Big Bang, the temperature of the Universe was over 10 trillion degrees Celsius. It was filled with energy in the form of photons – little "packets," or particles, of radiation.

The birth of matter

Under suitable conditions, high-energy photons can turn into particles of matter. And that is what happened in the early stages of the Universe. Most of the particles turned back into radiation. But some remained to form the atomic particles that make up the Universe as we know it today.

All of the subatomic particles – protons, neutrons, and electrons – had been formed by the time the Universe was 10 seconds old. From the instant it was created, the Universe began expanding and cooling. After approximately three minutes, its temperature had dropped to about one billion degrees. Then protons and

▶ When the Universe was created in the Big Bang, it was fantastically hot and filled with photons (particles of radiation). Heavy particles such as neutrons (n) and protons (p) formed when photons collided (A). But they were destroyed almost immediately, changing back into radiation (B). Likewise, most electrons (e) were destroyed (C). But a few protons, neutrons, and electrons survived (D). It was now just 10 seconds after the Big Bang. After about three minutes the protons and neutrons began coming together to form the nuclei of atoms. First came two forms of heavy hydrogen, called deuterium (E) and tritium (F). Finally came helium (G). Hydrogen and helium are the most plentiful elements in the Universe.

▼ The Earth was born about 4.6 billion years ago. Primitive life began about a billion or so years later. But early ancestors of the human race did not appear until just a few million years ago.

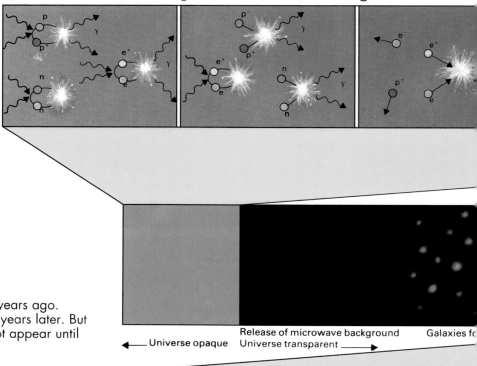

← Universe opaque   Release of microwave background / Universe transparent →   Galaxies fo

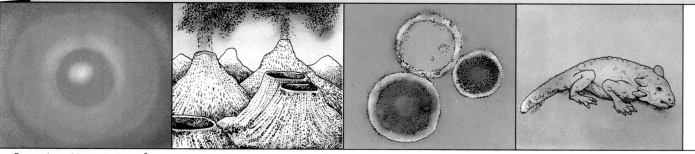

Formation of Earth (4.6×10^9)   Oldest terrestrial rocks (3.8×10^9)   Earliest life-forms (3.6×10^9)   First reptiles (3.0×10^8)

neutrons began combining to form the central cores, or nuclei, of atoms, such as helium. The Universe was then made up of radiation and matter, in the form of protons, helium nuclei, and electrons.

After several hundred thousand years the temperature had fallen to about 3,000°C (about 5,000°F). The protons were now able to capture and hold electrons. The protons became atoms of hydrogen, and the helium nuclei became helium atoms. With fewer particles about and the Universe greatly expanded, radiation could travel over vast distances without being either absorbed or deflected. And the Universe had become transparent.

Background radiation

Radiation has continued traveling since that time. It has been spreading out through larger and larger volumes of space as the Universe has continued to expand. Astronomers have worked out that such radiation would now give space an overall temperature of about -270°C (-454°F), about 3°C above absolute zero.

In 1965 two American scientists discovered that radiation at this temperature did indeed fill space. They were Arno Penzias and Robert Wilson. Their discovery of what is often called fireball radiation provided convincing evidence that the Big Bang theory is correct. The discovery also won them a Nobel Prize.

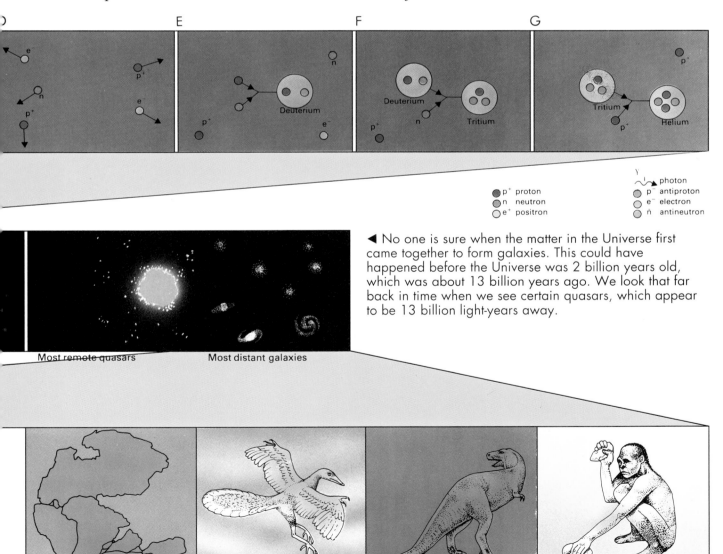

p⁺ proton		γ photon
n neutron		p⁻ antiproton
e⁺ positron		e⁻ electron
		ñ antineutron

◄ No one is sure when the matter in the Universe first came together to form galaxies. This could have happened before the Universe was 2 billion years old, which was about 13 billion years ago. We look that far back in time when we see certain quasars, which appear to be 13 billion light-years away.

Most remote quasars   Most distant galaxies

Separation of continents (2.0×10^8) First bird (2.0×10^8) Death of dinosaurs (6.5×10^7) First men $(2.5 \times 10^6$ years ago$)$

Open or closed?

Most astronomers are in agreement about how the Universe began with a Big Bang and how it has evolved since then. They are not so sure what will happen in the future. Certainly for many billions of years the Universe will continue to expand. But will it expand for ever?

Most astronomers think it will. They say we have an Open Universe. The only thing that could prevent the galaxies from flying apart forever would be gravity. For gravity to be powerful enough to do this, the Universe must have a certain mass. But there appears to be nowhere near enough mass in the stars and galaxies to halt the expansion.

Dark matter
However, astronomers know that stars and galaxies are not the only matter in the Universe. There is also dark matter, which we are largely unable to detect. There is dark matter in the dust clouds in space and in dead, burned-out stars. There is also matter hidden in the abyssal depths of black holes.

Other possible sources of dark matter are the atomic particles called neutrinos. Recent experiments have indicated that they might have a slight mass. If they have, they would greatly increase the density of the Universe because there are so many of them.

Crunch, bang, crunch . . .
If there is sufficient dark matter to halt the expansion of the Universe, then we have a closed Universe. Eventually the Universe will collapse in on itself and end in a Big Crunch. But things may not end here. Another Big Bang may be triggered off that will set the Universe expanding once again. In turn, there will be another Big Crunch, yet another Big Bang, and so on. We will have an oscillating Universe.

Fate of the Universe

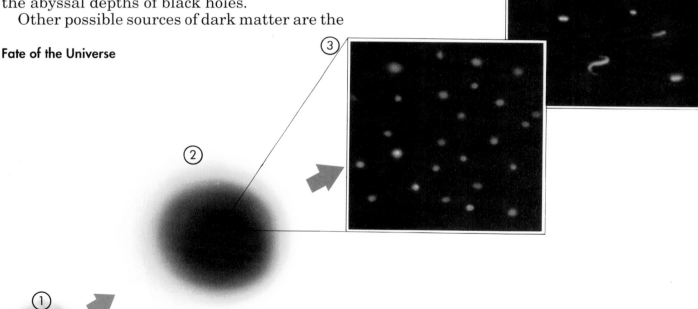

The pictures on this page show stages in the development of the Universe from the time of the Big Bang. Most astronomers are agreed on these stages (1–4).

1. The Universe is created, and time and space begin.
2. A dense opaque cloud of particles and radiation fills the cooling, expanding Universe.

3. The particles condense into galaxies, and stars begin to shine.
4. After 15 billion years, the Universe reaches the stage it is at today. It is filled with spiral, elliptical, and irregular galaxies rushing headlong through space, bringing about further expansion. What happens next will depend on whether the Universe is open or closed.

In an open Universe there is not enough matter present to hold back the outward-rushing galaxies. They will continue to move apart. Eventually they will run out of nuclear fuel and will start to fade. Maybe they will in time break down into particles and radiation.

Open Universe

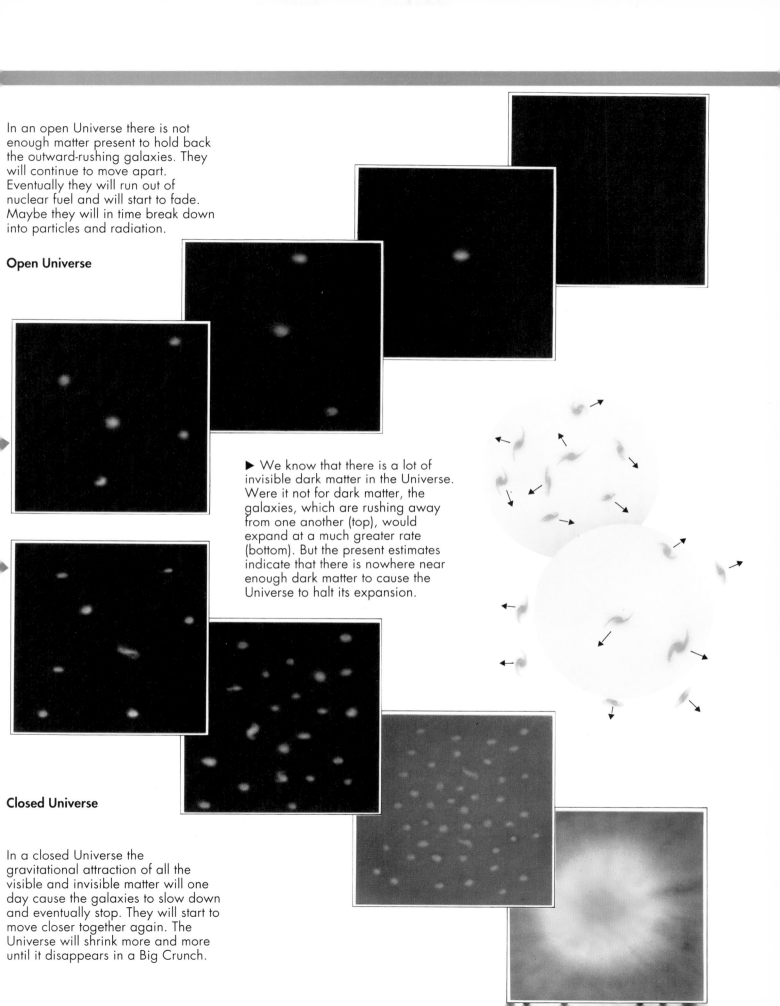

▶ We know that there is a lot of invisible dark matter in the Universe. Were it not for dark matter, the galaxies, which are rushing away from one another (top), would expand at a much greater rate (bottom). But the present estimates indicate that there is nowhere near enough dark matter to cause the Universe to halt its expansion.

Closed Universe

In a closed Universe the gravitational attraction of all the visible and invisible matter will one day cause the galaxies to slow down and eventually stop. They will start to move closer together again. The Universe will shrink more and more until it disappears in a Big Crunch.

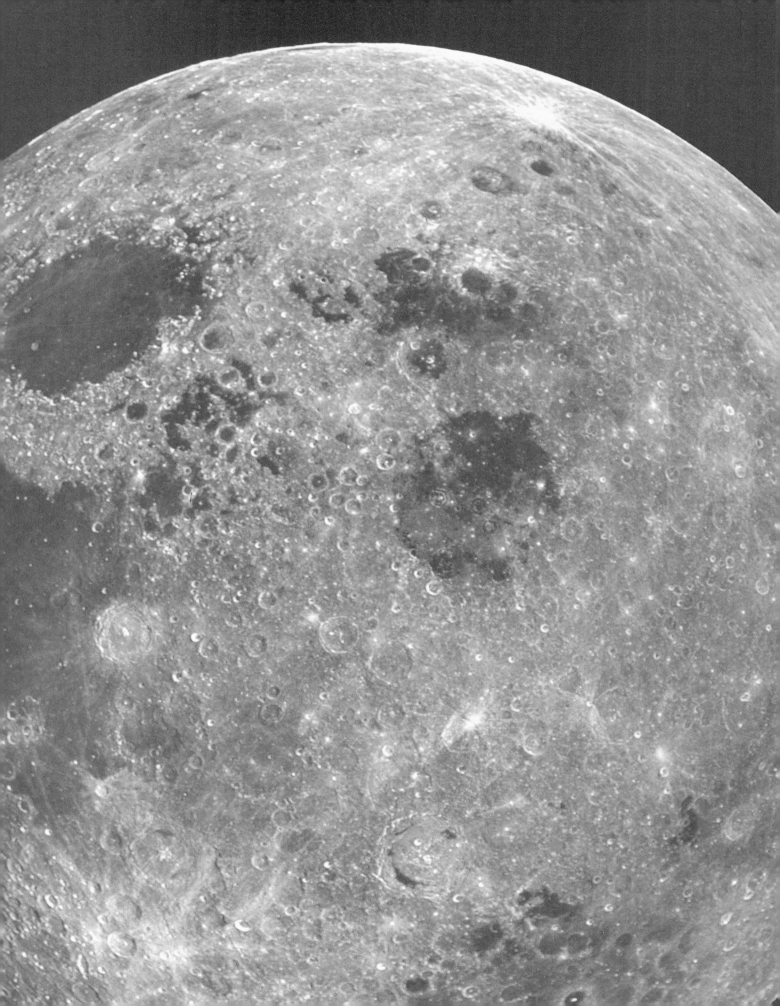

Part Two

The Solar system

Our home the Earth seems big to us, but it is a mere speck in a boundless Universe. Very much bigger is the Sun, the body that breathes life into our world. It is our star, similar to the other stars in the sky but very much closer.

The Sun holds the Earth captive in space by its enormous gravity. It also holds in check eight other large bodies, all of which, together with the Earth, we call the planets. The planets are the main members of the Sun's family, or Solar System. The family also includes many moons, rock lumps we call asteroids, and comets.

Since the Space Age began, our knowledge of the Solar System has expanded greatly. Space probes have voyaged to all but one of the planets, and sent back spectacular photographs of these far-off worlds, together with a wealth of fascinating information.

◀ The lunar surface, as viewed from the spacecraft *Apollo 8* in 1968. This was the first manned mission to orbit the Moon. It showed that the far side is more rugged than the near side.

The Solar System

Dominating our small corner of the Universe is the star we call the Sun. As it rushes headlong through space, it carries with it a collection of planets and moons, asteroids, meteoroids, and comets. Together they make up the Solar System. The heat and light pouring out from the Sun's nuclear furnace make the inner planets – Mercury and Venus – searingly hot and inhospitable. But the same heat and light breathes life into the planet Earth. The Sun was born along with the planets nearly 5 billion years ago from a cloud of interstellar gas and dust. It is now in middle age. It will be another 5 billion years before the Sun begins to run out of nuclear fuel, swells up into a red giant, and starts to die. When this happens, all life on Earth will perish.

SPOT FACTS

• A globe the size of the Sun could swallow more than 1 million Earths.

• The Sun contains 750 times more matter than all of the rest of the Solar System put together.

• Every second, the Sun loses more than 4 million metric tons of its mass. The lost mass is converted into vast amounts of energy in its nuclear furnace.

• It can take up to a million years for the energy produced in the Sun's core to reach the surface of the Sun.

• The temperature of the Sun's corona is about 1 million °C. It is still a mystery why it should be so hot, as it is heated by the solar surface at only 5,500°C.

• Sunspots occur where strong magnetic fields break through the Sun's surface. It is thought that these fields cause a flow of gas to the area, which cools the surface at that point.

• From 1645 to 1715, hardly any sunspots were observed. During this period, temperatures on Earth were far colder than usual, and the period is known as the "Little Ice Age." The link between sunspots and weather is not fully understood.

• The Sun is so bright that many people have been blinded by it. You should never look at it directly. It is believed that Galileo went blind because he observed the Sun through his telescope.

• The reactions that produce energy deep inside the Sun also release particles called neutrinos. So far, only a third of the predicted number of neutrinos have been observed. The reason for this is unknown.

• More than 70 chemical elements have been discovered in the Sun. The element helium was found there before it was detected on Earth. Helium derives its name from the Greek word for the Sun.

• The extreme edge of the Solar System – where the most distant comets are found – is thought to be about two light years from the Sun. This is halfway to the nearest star.

• Vibrations of the Sun's surface occur, in the same way that earthquakes vibrate the Earth.

• The Sun's rotation varies from 24.9 days at its equator to 34 days at its poles. Sunspots appear to take 27.5 days to return to the same position.

• Many searches for a tenth major planet have been made without success. More distant bodies than Pluto have been found, but they are all small.

• If you could stand on Pluto, the Sun would look no bigger in the sky than the planet Venus does from Earth, though it would still be 10,000 times brighter than the full Moon.

• If all the energy from the Sun could be collected and harnessed on Earth, just 1 sq cm (0.15 sq. in.) would receive enough energy to supply the power needs of a city with a population of some 30,000 inhabitants.

Our star, the Sun

To us on Earth the Sun is the most important star in the Universe. But as a star it is not very special. It appears bigger and brighter than the others only because it is much nearer – about 150 million km (93 million mi.) away. This compares with over 40 trillion km (25 trillion mi.) to the next nearest star.

Like most stars, the Sun is a globe of searing hot gas. It is an averaged-sized star, with a diameter of about 1,400,000 km (870,000 mi.), or more than 100 times that of the Earth. It gives off white light, and other kinds of energy, from short-wavelength gamma rays to long radio waves.

This energy is produced in the central core of the Sun at temperatures of up to 15 million degrees Celsius. There, nuclear reactions take place between nuclei of hydrogen. They combine, or fuse, together to form helium. During this process a little mass (m) is "lost," or rather converted into energy (E), according to Einstein's famous equation $E = mc^2$, where c is the velocity of light. The loss of only a small amount of mass releases huge amounts of energy.

The solar atmosphere

The energy produced in the Sun's core travels to the surface and then radiates into space. The surface, or photosphere, ("light sphere") has a temperature of about 5,500°C (10,000°F).

Above the photosphere is a layer of gases called the chromosphere ("color sphere"), about 10,000 km (6,000 mi.) thick. It is the inner part of the Sun's atmosphere, and is named after its reddish color. The outer atmosphere, the corona extends for millions of miles until it merges into space. It can be seen well only during a total eclipse of the Sun.

▶ This *Skylab* picture shows tongues of gas leaping thousands of miles above the surface of the Sun. The surface has been blotted out using an instrument called a coronagraph to create an artificial eclipse.

▼ Another *Skylab* picture shows in false colors the Sun's corona. It extends millions of miles out into space.

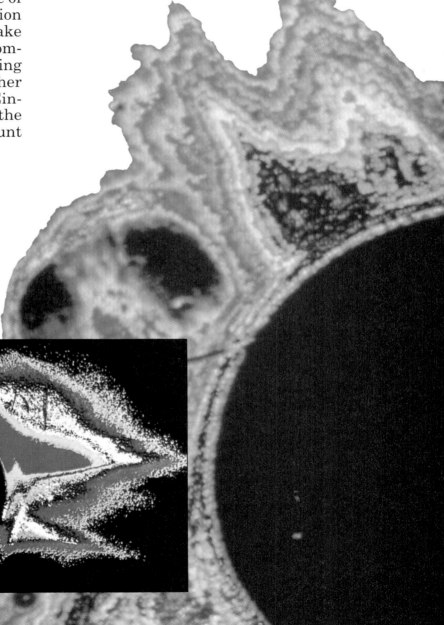

Spots and flares

The seething surface

When studied with astronomical instruments, the photosphere appears as a mass of speckles, or granules. They show where cells of hot gas are constantly welling up from below. Quite often, more prominent markings appear on the surface, called sunspots. They look rather like ink blots, with a dark middle, or umbra, surrounded by a paler penumbra. Sunspots are about 2,000°C (nearly 4,000°F) cooler than their surroundings.

Some sunspots, called pores, may be just a few hundred miles across and last for a day or so. Others may grow up to a width of 200,000 km (over 100,000 mi.) or more and last for months. The number of sunspots varies from year to year according to a regular cycle.

Sometimes when sunspots occur, brilliant and violent eruptions called flares take place. They give off powerful streams of charged particles, such as protons and electrons. When these particles reach Earth, they give rise to brilliant displays of aurora and disrupt radio communications. Some charged particles stream out from the Sun, forming the solar wind. On reaching Earth, they become trapped in its magnetic field to form "belts" of intense radiation, called the Van Allen belts.

▲ In far northern regions of the world the night sky is often lit by shimmering curtains of colored light. They are displays of the Northern Lights, or Aurora Borealis. Similar displays occur in far southern regions, where they are called the Southern Lights, or Aurora Australis. Aurorae happen when the solar wind blows strongly. It forces high-energy particles out of the Van Allen radiation belts. These collide with atoms in the upper atmosphere, which then give off an eerie light.

Eclipses

An eclipse of the Moon occurs when the Moon moves into the Earth's shadow in space. It moves first into a region of partial darkness, called the penumbra, then into a region of almost complete darkness, called the umbra. The Moon may remain in eclipse for up to about 2½ hours. During a lunar eclipse the Moon often takes on a faint glow. This is caused by sunlight reaching it after having been bent by the Earth's atmosphere. An eclipse of the Sun occurs when the New Moon moves in front of the Sun and blots out some or all of its light. In a partial eclipse only part of the Sun is blotted out. In an annular eclipse the Moon does not quite mask the Sun, leaving a ring (annulus) of light. In a total eclipse the Moon masks the Sun completely. A shadow, up to 250 km (150 mi.) wide, travels rapidly over the Earth's surface. Totality (time of total eclipse) lasts only up to about 7½ minutes.

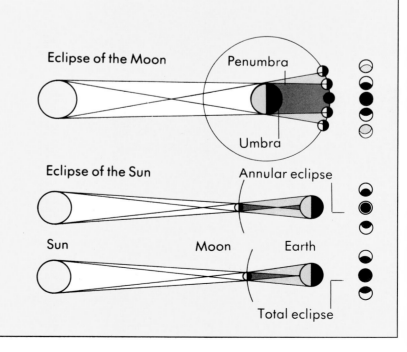

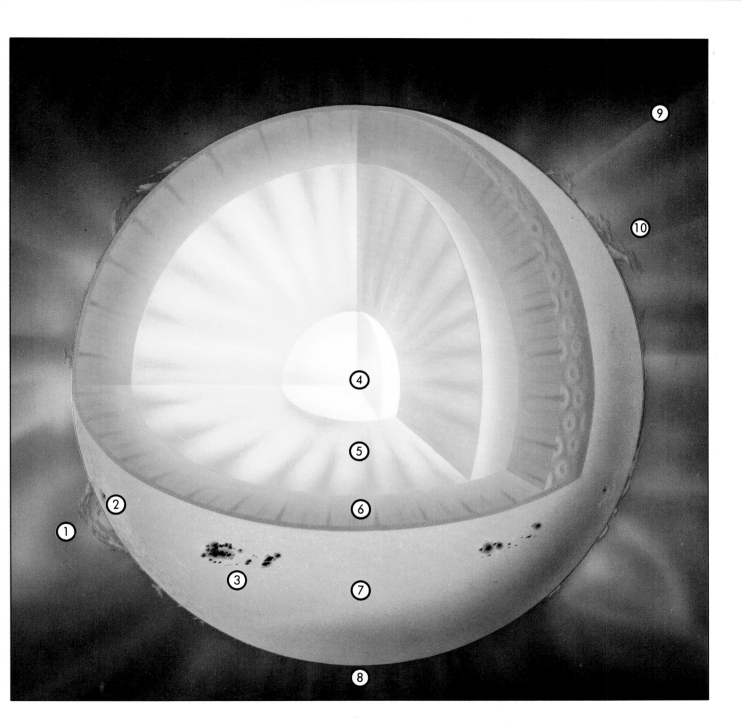

▲ In the central core of the Sun (4) the temperature is about 15 million degrees Celsius. At this temperature nuclear reactions take place which produce the energy that keeps the Sun shining. The energy travels toward the surface in two ways. On the first stage of its journey, it travels in the form of radiation, through the so-called radiative zone (5).

From the top of this zone the energy also travels by convection: hot gas rises and carries the heat to the surface. This region of the Sun is called the convective zone (6). From the surface, the photosphere (7), the energy radiates into space as light, heat, and other radiation. Sometimes dark patches called sunspots appear on the surface (3).

Sunspots can grow very large, and may be surrounded by bright patches called faculae (2). Just above the bright surface is a thin layer of atmosphere called the chromosphere (8). Through this layer, fountains of hot gas shoot up as filaments (1) or prominences (10). Farther out, the outer atmosphere, or corona (9), billows into space.

The planets

The Sun is over 300,000 times more massive than the Earth, and the pull of its gravity is very strong. It is this pull that keeps the Earth and eight other large bodies circling in space around the Sun. These bodies are the planets.

From Earth we can see with the naked eye five planets – Mercury, Venus, Mars, Jupiter, and Saturn. We need a telescope to see the other three planets – Uranus, Neptune, and Pluto.

The Earth itself is a planet. But until about 450 years ago, most people believed it was the center of the Universe. They thought that all the heavenly bodies – Sun, Moon, planets, and stars – revolved around the Earth. But this could not easily explain how the planets move through the heavens. At times, for example, they appear to loop backwards.

In the 1500s a Polish priest named Nicolaus Copernicus realized what was wrong. The Sun must be the center of everything, not the Earth. He put forward the idea of a Sun-centered system, or Solar System, in 1543, and in so doing gave birth to modern astronomy.

Worlds near and far

The Earth differs from the other planets in one very important respect. It boasts conditions that allow millions of different life-forms to flourish. On all the other planets, conditions are deadly to life as we know it.

The planets divide neatly into two groups. Mercury, Venus, and Mars are small rocky planets like the Earth. They are close enough to be considered neighbors. Jupiter, Saturn, Uranus, and Neptune on the other hand are giant balls of gas. Like icy Pluto, they are far-distant worlds. The diagram at the foot of these pages shows the different sizes of the planets and compares them with the Sun.

All of the planets except Mercury and Venus are the center of miniature systems of their own. They have moons circling around them. The Earth has only one Moon, but the giant planets have a multitude. Details of the numbers of moons, and other information about planetary sizes and orbits, are given in the table on the opposite page.

The remaining members of the Sun's family are much smaller. They include the asteroids – miniature planets that circle the Sun between the orbits of Mars and Jupiter – as well as meteoroids and comets. Meteoroids that enter the Earth's atmosphere and burn up flash across the night sky as meteors, or shooting stars. A very few reach the ground as meteorites. All of these smaller bodies make up what is called the debris of the Solar System.

Jupiter

Mercury　　Venus　　　　Earth　　　Mars

Sun

Planetary statistics

Planet	Mercury	Venus	Earth	Mars	Jupiter	Saturn	Uranus	Neptune	Pluto
Diameter at equator (km)	4,878	12,104	12,756	6,794	142,800	120,000	51,800	49,500	2,300
Mean distance (million km)	59.9	108.2	149.6	227.9	778.3	1,427.0	2,869.6	4,496.7	5,900
Mean distance (Earth = 1)	0.387	0.723	1.000	1.524	5.203	9.539	19.182	30.058	39.44
Circles Sun in (d, y)	87.97d	224.7d	365.2d	686.98d	11.86y	29.46y	84.01y	164.79y	247.7y
Turns on axis in (h, d)	58.65d	243d	23.93h	24.62h	9.8h	10.2h	17.2h	16.1h	6.4d
Mass (Earth = 1)	0.056	0.815	1.000	0.107	318	95.1	14.5	17.2	0.002
Volume (Earth = 1)	0.05	0.88	1.00	0.15	1.316	755	52	44	0.005
Density (Water = 1)	5.43	5.24	5.52	3.04	1.32	0.70	1.27	1.77	2
Number of moons	0	0	1	2	16+	22+	15+	8	1

▶ The orbits of the planets in the Solar System, seen from "above." Most are nearly circular, with the Sun at the center. Those of Mercury and Pluto are displaced off-center, and Pluto's orbit sometimes crosses inside that of Neptune.

▼ The sizes of the planets drawn to the same scale. From left to right, they are shown in order of increasing distance from the Sun. Jupiter, Saturn, Uranus, and Neptune dwarf the others, but are themselves dwarfed by the Sun.

Saturn

Uranus

Neptune

Pluto

Planets nearby

Mercury, Venus, and Mars are close enough to Earth to be considered neighbors. But they are strikingly different from Earth in most respects. Mercury is a cratered wasteland, where noonday temperatures soar high enough to melt lead. It is a small planet, a little larger than our Moon, and is the planet nearest to the Sun. Earth's near-twin Venus is a hellish world that is as hot as Mercury and has a dense, suffocating atmosphere. It is permanently covered in clouds. Mars, on the other hand, is icy cold, with only a whiff of an atmosphere around it. But it is the only planet in the Solar System where human beings could survive.

SPOT FACTS

• Mercury travels faster in its orbit than any other planet. Its average speed is over 170,000 km/h (over 100,000 mph), or 1.5 times the speed of the Earth.

• The daytime and night-time surface temperature of Mercury has the widest-known range in the entire Solar System.

• The extinct volcano Olympus Mons on Mars is nearly three times the height of Mount Everest.

• In 1877 the Italian astronomer Giovanni Schiaparelli said he could see *canali* (channels) on Mars. This was translated as "canals," prompting people to think that they were artificial waterways made by intelligent beings.

• The greenhouse effect makes the surface of Venus even hotter than that of Mercury. But the cloud cover means that the temperature varies by only 10°C between day and night.

• There are 100,000 volcanoes on Venus, though it is not certain that any are still active.

• Almost all the features on Venus are named for women – Venus is the only planet that shares a name with an ancient goddess (of love) rather than a god.

• The surface of Venus is the youngest of any of the planets at 400 million years. Lava flows from the numerous volcanoes have reshaped its surface many times over.

• Venus may once have had water on its surface, but the greenhouse effect has increased its temperature so much that now no water exists. Some scientists are worried that the same thing may happen on Earth.

• The clouds of Venus prevent much light reaching the surface. The light is orange and similar to the brightness of an overcast winter's day on Earth.

• Nine space probes have landed on Venus. Not one lasted more than a few hours under the planet's extreme pressure and temperature.

• The Mariner Valley on Mars is so big that it could contain the entire Rocky Mountains.

• In the distant past, when there may have been oceans on Mars, life could have formed. Scientists believe that fossil evidence for primitive bacteria may still be found on Mars.

• Winds on Mars can reach 96 km/h (60 mph) blowing the fine dust into huge sand-dune systems.

• Changes in the appearance of the dark areas on Mars, once thought to be caused by life forms, are now attributed to the movement of wind-blown dust.

• Occasionally there are giant dust storms on Mars that can cover the entire planet, blotting out the Sun for weeks.

• Samples from Mars have already arrived on Earth in the form of meteorites. They were blasted from the Martian surface by impacts of space debris over billions of years.

Mercury

Mercury orbits so close to the Sun that it always appears near the Sun in the sky. We can therefore see the planet only in the morning sky in the east just before sunrise or in the evening sky in the west just after sunset. At its brightest Mercury looks like a bright pinkish star. Through a telescope, you can make out little detail on the planet's disk, except for a few vague markings.

Mercury is the second-smallest planet, after Pluto. It is not much bigger than the Moon and, like the Moon, does not have a substantial atmosphere. It looks remarkably like the Moon too, because it is covered with craters. These were formed billions of years ago when the planet was bombarded by huge meteorites.

One difference between Mercury and the Moon is that it does not have such large "seas," or maria. Its most obvious feature is the huge, ring-shaped Caloris Basin, which measures some 1,400 km (900 mi.) across. The basin must have been carved out by the impact of a truly gigantic meteorite. The same impact caused the surface to wrinkle over vast distances, forming "waves" of mountain chains.

▲ This photograph of Mercury was taken by *Mariner 10* when it flew past the planet in 1974. The surface looks very similar to that of the Moon. It is peppered with craters, large and small. Some craters have bright rays radiating from them. Others have mountain peaks in their center. Both these features are common also on the Moon. In places the surface of Mercury is quite smooth, with few craters. But there are no large "seas" like there are on the Moon.

▶ The structure of Mercury. Astronomers think that it has a large core of iron and nickel. Above the core there is probably a layer of lighter rock, which is topped by an even lighter crust.

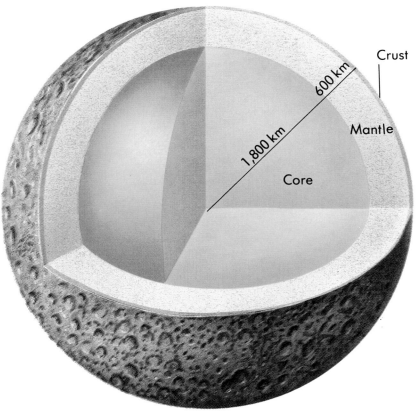

Crust

600 km

Mantle

1,800 km

Core

Venus

Venus is the planet that comes closest to Earth within 42 million km (26 million mi.). In size it is almost an identical twin of the Earth. Yet the two planets are very different in other ways. Venus rotates very slowly on its axis; its day is 243 Earth-days long. It also rotates in the opposite direction from the Earth and indeed from most of the other planets. Apart from the Moon, Venus is by far the brightest object in the night sky. Sometimes we see it in the west at sunset, and call it the evening star. At other times we see it in the east at sunrise, and call it the morning star. Seen through a telescope, Venus shows phases like the Moon. Its shape changes from crescent to full and back again.

Conditions on Venus are quite different from those on Earth. The temperature at the surface is over 480°C (900°F). The atmospheric pressure is nearly 100 times that on Earth. The reason for these hellish conditions is the thick atmosphere. It is made up mainly of carbon dioxide. This is a heavy gas, which traps heat in the same way as a greenhouse.

The atmosphere of Venus is always full of clouds. They reflect sunlight brilliantly, which is what makes Venus so bright. The clouds are made of tiny droplets of sulfuric acid. The acid was probably formed from sulfur dioxide gas shot into the atmosphere when volcanoes erupted on the surface.

◄ This three-dimensional view of the landscape of Venus was made by the *Magellan* space probe that orbited Venus between 1990 and 1994. It used radar to penetrate the clouds, and the results were converted into realistic images, although the height of features was exaggerated by 20 times. This picture shows the Maat Mons volcano; it rises 8 km (5 mi.) high. In the foreground are the lava flows that extend for hundreds of miles across Venus's fractured plains.

▼ A close-up of the surface of Venus. Despite the crushing pressure and intense heat, space probes have landed on the planet to take color photographs. This one was taken by the Soviet *Venera 13*.

Beneath the thick clouds on Venus, the surface consists mainly of vast rolling plains. Here and there are isolated lowlands and highlands. The plains are covered by flows of lava, or solidified rock, that have poured from the planet's 100,000 volcanoes. Some of these may still be active. There are very few meteorite craters, as found on Mercury, because the surface is often covered over by lava flows.

There are two main highland regions, or continents. The one in the northern hemisphere is called Ishtar Terra. It is about the size of Australia. The other continent, the larger of the two, lies near the equator. It is called Aphrodite Terra, and is about the size of Africa.

▼ The structure of Venus. Venus probably has much the same structure as the Earth. It has a central core, a thick mantle, and a thick crust. The core of iron and nickel may be partly molten. Above this lie a rocky mantle and the surface crust. The crust is more than twice as thick as that of the Earth.

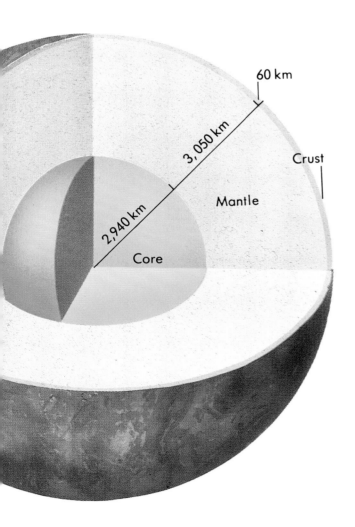

60 km

3,050 km

2,940 km

Crust

Mantle

Core

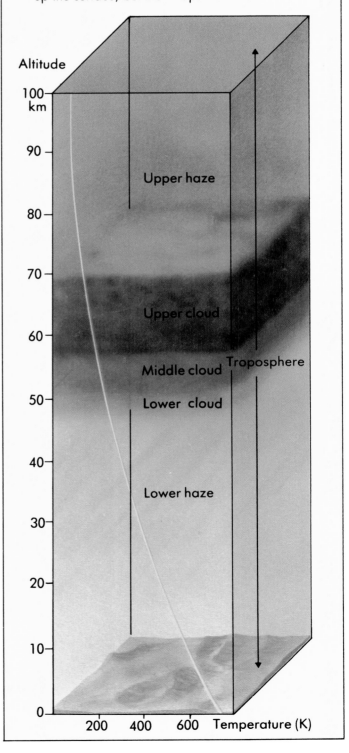

Venus's atmosphere

The atmosphere of Venus is made up mainly of carbon dioxide (96 percent) and nitrogen. The heavy atmosphere lets in solar energy to heat up the surface, but then traps the heat.

Altitude

100 km

90

80

70

60

50

40

30

20

10

0

Upper haze

Upper cloud

Middle cloud

Lower cloud

Lower haze

Troposphere

200   400   600   Temperature (K)

Mars

The structure of Mars

Mars is smaller than the Earth. It has a diameter of 6,794 km (4,222 mi.) at the equator – just over half that of the Earth. It is also thought to have a layered structure like the Earth. At the center is a core of iron and iron compounds. The core is surrounded by a deep mantle of silicate rock, and on top is a relatively thick crust. The crust is pitted by craters where it has been bombarded with rocks from outer space. The Martian surface gets its rusty red color from iron oxides in the soil. The highest mountain is Olympus Mons (bottom), with a height of 25 km (16 mi.). It is far bigger than Earth's highest peak, Mount Everest.

Mars comes nearer to Earth than any other planet except Venus. Close approaches occur about every 26 months at the time of opposition. At this time, the two planets orbit for a while side by side. At the closest oppositions, Mars is less than 56 million km (35 million mi.) away.

Of all the planets, Mars is the easiest to recognize in the night sky. It shines with a distinct red-orange glow. For this reason it is often called the Red Planet.

In some respects Mars is similar to Earth. A day on Mars is only 40 minutes longer than a day on Earth. Mars also has seasons. This is because its axis – like Earth's – is tilted at an angle to the plane of its orbit. But the Martian year (687 Earth-days) is almost twice as long as Earth's, so the seasons are nearly twice as long too. Mars is also similar in having ice caps at the poles. These caps vary in size from season

▶ Ice clouds on Mars. As the Sun rises over canyons on Mars, its rays enable puffy clouds of ice crystals to be seen. This region of Mars has the delightful name of Labyrinth of the Night (Noctis Labyrinthus).

▼ A *Viking* probe sent back this photograph as it approached Mars in 1976. Three interesting features stand out. To the left is one of Mars's four huge volcanoes, Ascraeus Mons; in the middle is the great scar of Mariner Valley; and to the right is the Argyre Basin, covered in frost.

to season. They shrink in the spring and grow again in the fall.

The wave of darkening

Mars is not covered with dense clouds like Venus. And through a telescope, we can often see various features on its surface. There are dark markings that rotate with the planet. One of the best known, near the Martian equator, is named Syrtis Major.

There are other dark markings that change with the seasons. As the ice caps shrink at the north and south poles, a so-called wave of darkening sweeps from them toward the equator. Some people once thought that this effect might be caused by the growth of vegetation, fed by water melting from the ice caps. But space probes have shown there are no plants, nor life of any kind, on Mars.

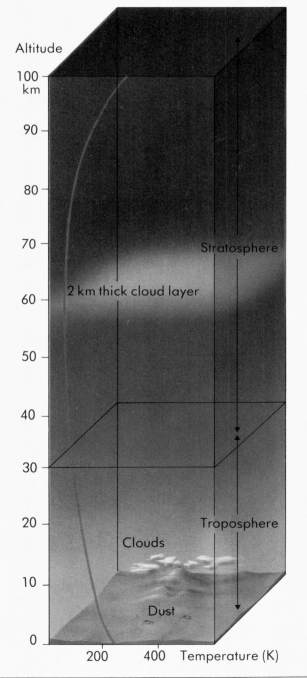

The Martian atmosphere

The atmosphere on Mars is mostly carbon dioxide at a pressure less than one-hundredth that of Earth. In the lower part of the atmosphere clouds of water ice often form. Dust is always present, giving the Martian sky a pinkish color. The temperature (red line) changes the higher you go above the surface.

Altitude

100 km

90

80

70 — Stratosphere

2 km thick cloud layer

60

50

40

30

20 — Troposphere

Clouds

10

Dust

0

200   400   Temperature (K)

The Martian landscape

We can see few details of the Martian landscape through a telescope. Most of our information has come from space probes, such as *Mariner* and *Viking*. They have photographed the planet both from orbit and from the surface.

The northern hemisphere consists mainly of low-lying plains with relatively few craters. The southern hemisphere has a more ancient crust which is heavily cratered. It also has two huge basins, gouged out millions of years ago by the impacts of massive meteorites. The largest basin, Hellas, is more than 1,600km (1,000 mi.) across. It is twice as big as the Argyre Basin.

One of the most interesting Martian features is a great gash in the surface that runs for some 5,000 km (roughly 3,000 mi.) near the equator. In places it is more than 200 km (120 mi.) wide and 5 km (3 mi.) deep. It is called Mariner Valley (Valles Marineris).

Northwest of Mariner Valley are four huge extinct volcanoes. Three stand in a row on the Tharsis Ridge and soar 20 km (over 10 mi.) high. They are dwarfed by the fourth, Olympus Mons. It is 5 km (3 mi.) higher and rises from a base approximately 600 km (nearly 400 mi.) across.

Rivers on Mars

Around these and other volcanoes are channels where molten lava once flowed. Elsewhere, there are other channels that look remarkably like dried-up river beds on Earth. So did rivers once flow on Mars? It is almost certain that they did, many millions of years ago.

Water probably flowed when the volcanoes were erupting. Water vapor and gases would have been given off during the eruptions. This would have created quite a dense atmosphere, allowing clouds to form and rain to fall. No rivers flow on Mars today, but there are still traces of water vapor in the atmosphere. This gives rise to the occasional cloud and early morning mist in the canyons. Astronomers believe that water is also locked in the ice caps at the two poles.

▼ This map of the landscape of Mars was produced from radar scans made from Earth. The most striking feature is the Tharsis Ridge (centered on longitude 90°). It is topped by three huge volcanoes. To the southeast is the circular depression of the Argyre Basin. Farther east is a much bigger basin called Hellas.

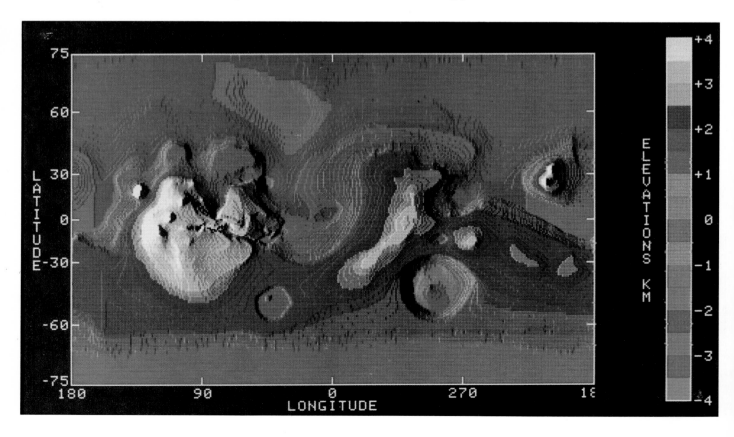

► Ice on Mars. Large patches of ice still linger on the rusty red ground of Mars. This picture was taken in midsummer near the north pole. The ice patches are all that is left of the ice cap that formed the previous winter. Ice caps advance from both poles toward lower latitudes during the winter. They shrink as the ice evaporates during the following summer. The ice cap grows largest at the north pole. The ice caps are made up mainly of water ice, together with small amounts of dry ice, or solid carbon dioxide.

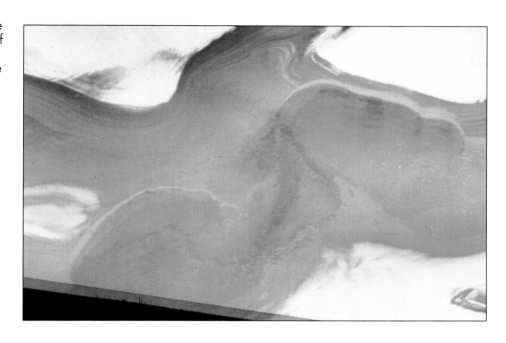

Vikings on Mars

In 1976 two American *Viking* space probes went into orbit around Mars. They then released craft that landed on the surface to take photographs. On the right is a view from *Viking 1* and below is a view from *Viking 2*. The pictures show a remarkably similar landscape. The ground is covered with fine rust-colored soil, and rocks are strewn everywhere. They are pitted by the impact of soil particles blown by the wind.

Far-distant worlds

Five frozen planets orbit in the outer reaches of the solar system, hundreds of millions of miles away. Four of them – Jupiter, Saturn, Uranus, and Neptune – are giant gas balls of enormous dimensions. Even though these planets are so far away, we now know a lot about them. Space probes have visited them and sent back closeup photographs and other information. Only the tiny ice world of Pluto remains almost a complete mystery.

SPOT FACTS

• Jupiter has more than twice the mass of all the other planets put together.

• Jupiter and the other three giant planets have no solid surfaces but are covered by an ocean of liquid hydrogen.

• Jupiter, Saturn, and Neptune give out twice as much heat as they receive from the Sun.

• Jupiter's rings are so faint that even if you were in a spacecraft nearby you could hardly see them. The particles in them are tiny, like grains of flour.

• In 1995 an atmospheric probe was released into Jupiter's atmosphere from the space probe *Galileo*. It was buffeted by winds of up to 643 km/h (400 mph).

• The *Galileo* probe also showed that there is less lightning in Jupiter's atmosphere than expected – about a tenth the amount that occurs on Earth. But when lightning does strike, it has ten times the force of lightning on Earth.

• Jupiter's Great Red Spot has been observed to absorb other smaller spots, adding their energy to its own to keep it rotating.

• Observations in 1989 from *Voyager 2* showed that Neptune's atmosphere had a great spot, similar to that on Jupiter. Photographs from the Hubble Space Telescope showed that by 1995 this had disappeared.

• At 15-year intervals, Saturn's rings appear edge-on, and astromomers look for satellites that would otherwise be hidden in the glare of the rings.

• A great storm appears on the surface of Saturn roughly every 30 years, in the form of a white spot. The reasons for its appearance are unknown.

• Saturn has the strongest winds in the Solar System. At the equator, they blow at up to 1,800 km/h (1,100 mph).

• The rings of Saturn cannot last for more than a few tens of millions of years unless they are renewed somehow. We may be seeing them at their most splendid, and in a few million years they may be gone.

• The spokes that are visible in Saturn's rings are believed to be caused by magnetic fields influencing tiny ice crystals.

• Although one of the poles of Uranus faces away from the Sun for many years at a time, measurements from spacecraft show that it is as warm as the sunlit pole.

• Pluto's orbit occasionally crosses that of Neptune. But because its orbit is at a different angle from Neptune's there is no danger of a collision.

• Since 1979, Neptune has been the most distant planet, and it will remain so until 1999. In that year, Pluto will move outside Neptune's orbit, and resume its role as the most distant planet.

• Pluto currently has a very thin atmosphere of methane. Due to its elliptical orbit, this probably freezes completely when the planet is at its most distant from the Sun.

• In 1994 a comet named Shoemaker-Levy 9 collided with Jupiter, leaving dark spots in the cloud decks that remained for months.

Jupiter

Beyond the Red Planet Mars, the next planet as we move away from the Sun is Jupiter. Between Mars and Jupiter there is a gulf of over 500 million km (300 million mi.). There we might perhaps expect to find another planet. But instead there is a collection of small rocky bodies, orbiting together in a broad band, or belt. They are the asteroids, or minor planets.

Apart from the Moon and Venus, Jupiter is the most brilliant object in the night sky. It shines so brightly because of its huge size and its cloudy atmosphere, which reflects sunlight very well. In Roman mythology Jupiter was king of the gods. It is an appropriate name for the largest planet, which is big enough to swallow 1,300 planets the size of the Earth. Its diameter is 11 times that of Earth.

In some respects Jupiter is more like a star than a planet like the Earth. Like a star, it is made up of gas, mainly hydrogen and helium. It has a powerful magnetic field, and gives off heat, radio waves, and even X-rays. Without doubt, if Jupiter had been much bigger, it would have begun to shine as a star.

In common with the other giant planets, Jupiter is the center of a large satellite system. At least 16 moons circle around it. The largest, Ganymede, is bigger than the planet Mercury. And like Saturn, Uranus, and Neptune, Jupiter has rings around it. But they are too faint to be seen from Earth.

When you look at Jupiter through a powerful telescope, you can see that the disk is banded with light and dark, reddish stripes. The light ones are called zones; the dark ones, belts. Within the zones and belts are all kinds of contrasting patterns described as spots, ovals, streaks, wisps, and plumes.

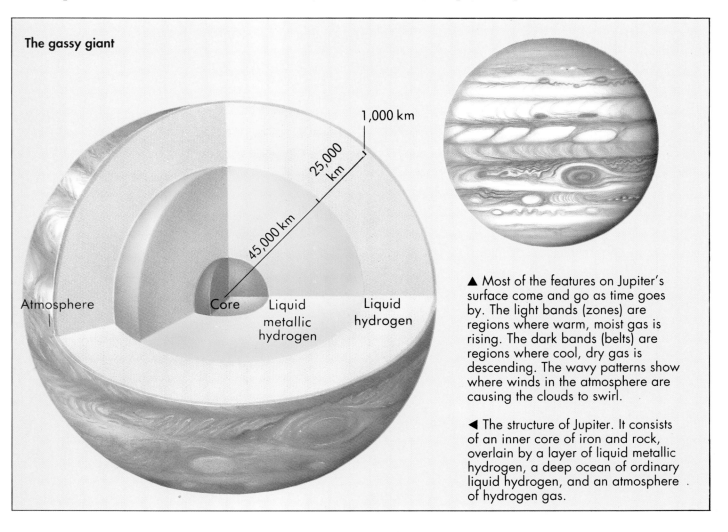

The gassy giant

1,000 km

25,000 km

45,000 km

Atmosphere

Core

Liquid metallic hydrogen

Liquid hydrogen

▲ Most of the features on Jupiter's surface come and go as time goes by. The light bands (zones) are regions where warm, moist gas is rising. The dark bands (belts) are regions where cool, dry gas is descending. The wavy patterns show where winds in the atmosphere are causing the clouds to swirl.

◄ The structure of Jupiter. It consists of an inner core of iron and rock, overlain by a layer of liquid metallic hydrogen, a deep ocean of ordinary liquid hydrogen, and an atmosphere of hydrogen gas.

Jupiter's weather

Jupiter rotates on its axis very quickly. It has the fastest rate of spin of all the planets, rotating once in less than 10 hours. This produces several noticeable effects. It causes the planet to bulge at the equator, and to be flattened at the poles. It also causes the clouds in the atmosphere to be drawn into alternate light and dark bands that run parallel with the equator (the belts and zones).

The rapid rotation sets up powerful wind belts, or jet streams, which blow east to west. They combine with rising and descending currents to churn up the atmosphere. This churning creates the waves, eddies, and other features we see on the disk. The biggest and most long-lived feature is the Great Red Spot. It is a huge storm in the atmosphere.

The white clouds in the atmosphere are made up of crystals of ammonia ice. They are higher and cooler than the reddish clouds, which are probably a compound of ammonia and hydrogen sulfide. Lower still, the clouds appear bluish and are probably water-ice crystals.

Space probes have spotted other kinds of activity in the atmosphere. They have observed lightning flashes during Jupiter's night, and also displays of aurorae in the polar regions. These displays are like the Northern Lights on Earth but on a much larger scale. They occur when charged particles from Jupiter's radiation belts collide with atoms in the atmosphere. The radiation belts are regions of space in which particles have become trapped by Jupiter's powerful magnetic field.

The Great Red Spot

Astronomers first noticed Jupiter's Great Red Spot about 300 years ago. And it has been visible for most of the time since then. It has varied in size over the years and at present measures about 28,000 km (over 17,000 mi.) long and 14,000 km (8,700 mi.) across. The Red Spot lies in Jupiter's southern hemisphere, and remains at a latitude of about 20°. Astronomers have had various theories about it. But space probes have shown it to be a great whirling storm center. It spins counterclockwise, making a complete rotation every 6 days or so. Scientists think that its color is caused by the presence of red phosphorus.

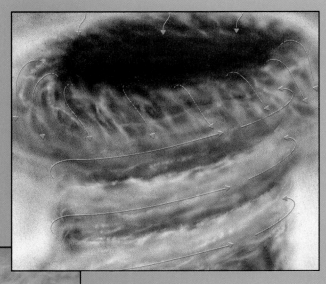

▲ The Great Red Spot appears to be a region of high pressure. The top stands about 8 km (5 mi.) higher than the surrounding cloud layer.

◀ The *Voyager 1* space probe took this close-up picture of the Great Red Spot. It shows great swirls in the atmosphere above and two whitish spots below. They are storm centers.

▶ Jupiter as taken by the Hubble Space Telescope in 1994. The impacts of fragments of the comet Shoemaker-Levy 9 caused a line of dark spots. The Great Red Spot is also visible.

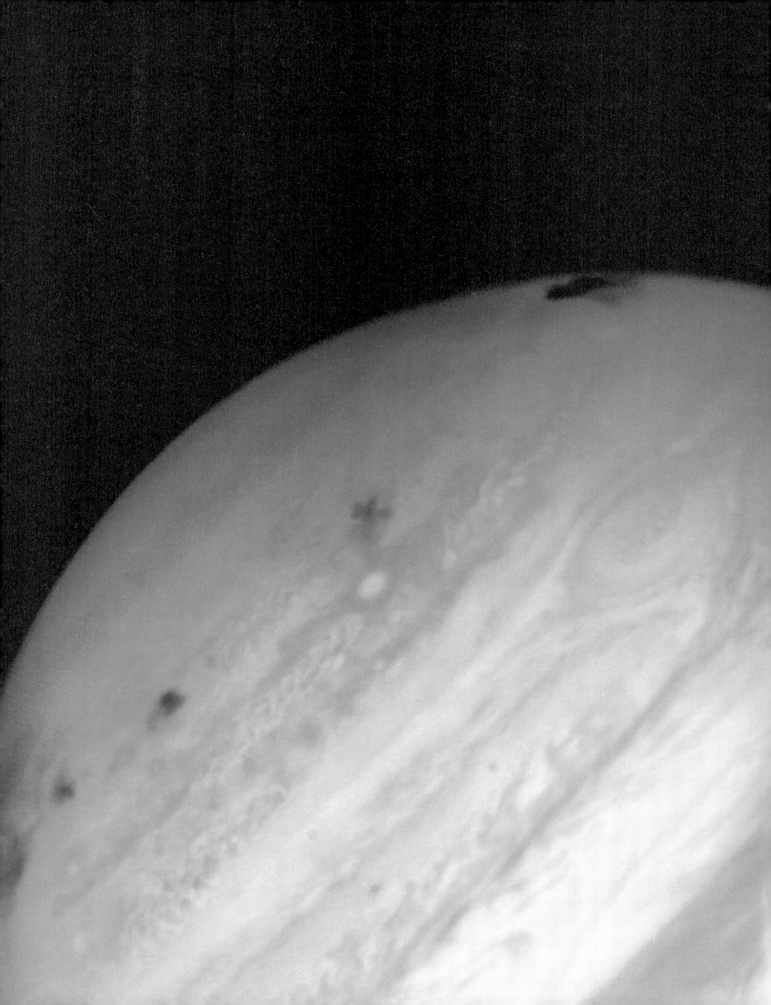

Saturn

Saturn was regarded as the outermost planet known until the late 1700s. It is nearly twice as far away from the Sun as Jupiter, at a distance of over 1.4 billion km (nearly 900 million mi.). It takes nearly 30 Earth-years to make one complete journey around the Sun.

Saturn is easily visible to the naked eye. But we need a telescope to see its most remarkable feature, its system of flat shining rings. From one side to the other the rings measure over 270,000 km (170,000 mi.) and they are about 60,000 km (40,000 mi.) wide. They reflect light well, and Saturn would be less visible and much fainter without them. We see different aspects of them as the planet travels in its orbit.

Floating on water

After Jupiter, Saturn is the largest planet, with a diameter nearly 10 times that of Earth. Apart from its prominent rings, it resembles Jupiter in many respects. It is a gas giant, surrounded by many moons (at least 22). It rotates rapidly on its axis, bulges at the equator, and is flattened at the poles. We can see on photographs of the planet's disk parallel belts and zones, which are bands of circulating clouds. But they are not nearly as marked or as colorful as Jupiter's are. Saturn also has a powerful magnetic field and is surrounded by radiation belts similar to the Van Allen belts around the Earth.

Saturn has a much lower relative density than Jupiter, only 0.7. This is the lowest of all the planets and is less than the density of water, so if Saturn could be placed in a vast ocean of water, it would float.

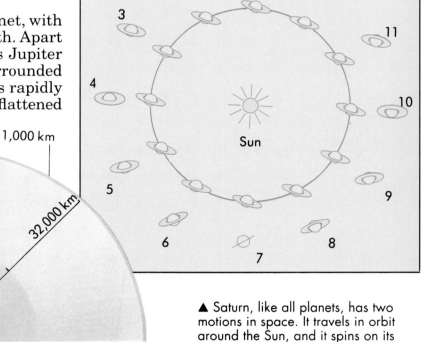

1,000 km

32,000 km

28,000 km

Core

Liquid hydrogen metallic

Liquid hydrogen

Atmosphere

▲ Saturn, like all planets, has two motions in space. It travels in orbit around the Sun, and it spins on its axis. The rings, which girdle the equator, present different aspects during the 29 or so years it takes Saturn to orbit the Sun. Sometimes we can see the rings edge-on (1, 7).

◄ Saturn, like Jupiter, probably has a small core of rock and iron with perhaps 3–10 times the mass of the Earth. Around the core is a thick layer of hydrogen in a highly compressed metallic form. Next comes a deep ocean of ordinary liquid hydrogen. The atmosphere is mainly hydrogen, with a little helium.

▲ Some of the wind belts in the north of Saturn. False colors have been used in this picture to bring out extra detail. The wavelike features and the oval "whirlpools" show where the atmosphere is in vigorous motion.

▼ A picture of Saturn taken by the Hubble Space Telescope in December 1994, showing a rare white spot that appeared on the planet. This was caused by white ammonia crystals forming above an upward flow of gas.

Speedy winds

Saturn is a very windy planet. Mostly the winds blow from west to east. They are strongest around the equator, where they reach the phenomenal speed of 1,800 km/h (1,100 mph). Wind speeds fall off going northward and southward away from the equator.

After a latitude of about 35° north and south of the equator, a curious thing happens. The winds suddenly change direction and start blowing the opposite way, from east to west. At higher latitudes still, the winds reverse direction again, and alternate like this all the way to the poles. The winds therefore form a series of bands, or belts.

Great churning, or turbulence, occurs at the boundaries of these alternating wind belts. This is accompanied by huge storms which rage there. They appear on the disk as pale or dark spots. But none of them is as large, or lasts so long, as the Great Red Spot on Jupiter.

Rings and ringlets

The bright shining rings that girdle Saturn are one of the wonders of the Solar System. Jupiter, Uranus, and Neptune have rings too, but they cannot be seen from Earth. Saturn's rings, however, present a magnificent sight in a telescope. They span a distance of more than 270,000 km (170,000 mi.).

In a telescope we can see three main rings, called A, B, and C. The outermost A-ring is separated from the B-ring by a dark gap, called the Cassini division. Both rings are bright. Inside the B-ring is the much fainter C-ring, also called the crepe ring. The rings, made up of particles of rock and ice, are broad but thin. When they are viewed edge-on, they almost disappear from view. In places they are less than 200 m (600 ft.) thick.

It was once thought that the rings were the remains of a close satellite of Saturn. But it is more likely that they are made up of material left over when the planet was formed.

Racing ringlets

When the *Voyager* space probes visited Saturn, the rings proved more fascinating than ever. They turn out to be made up of thousands of individual ringlets. The ringlets mark the path of particles of ice and rock as they race swiftly around the planet's equator. The particles vary in size from tiny grains to boulders more than 10 m (30 ft.) across.

The *Voyager* probes also discovered several new rings, which we cannot see from Earth. Inside the C-ring is a very faint D-ring, which probably extends down to Saturn's cloud tops. There are also other rings outside the A-ring. The most interesting is the very narrow F-ring, which is a curiously twisted collection of several ringlets. The probes also revealed a number of tiny moons orbiting close to the edges of some of the rings. They are called shepherd moons because they appear to help to keep particles within the rings.

▲ This picture shows clearly the A- and B-rings of Saturn, separated by the dark "gap" of the Cassini division. In the B-ring are curious dark, fingerlike features called spokes, which rotate with the ring.

◄ The ringlets that make up Saturn's rings. This picture has been computer-processed to show the ringlets in false color. Mostly it shows the C-ring, with the B-ring on the left. The different colors indicate that the rings are made up of different kinds of particles.

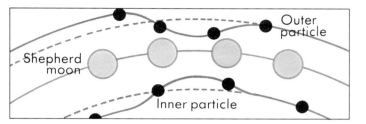

▲ Astronomers think that tiny shepherd moons may help to produce the ringlets. As a moon moves through the ring, a faster inner particle is slowed down by the moon's gravity and falls to a lower orbit. A slower outer particle, on the other hand, is speeded up and will then move to a higher orbit.

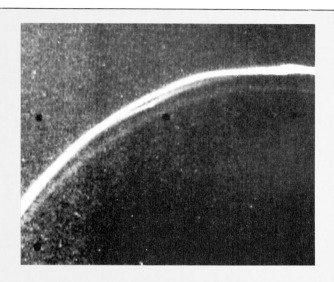

The fascinating F-ring

Outside Saturn's A-ring is the F-ring, which is less than 150 km (90 mi.) wide. It is far too thin to be seen from Earth. It is made up of as many as 10 ringlets. The ringlets are intertwined, giving a braided appearance (above). It is thought that the braiding effects are caused by the gravitational disturbance of a pair of nearby shepherd moons.

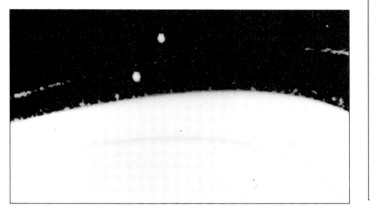

▶ The *Voyager 2* probe took this beautiful picture of Saturn from a distance of 3.4 million km (2.1 million mi.) as it looped away from the planet to its next port of call, Uranus. The picture reveals the classic A-, B-, and C-rings. Notice how transparent the inner ring is. The B-ring is denser and blots out the planet's disk. Note also the dark shadow cast on Saturn by the rings.

Outer planets

The English astronomer William Herschel discovered Uranus in 1781, after first thinking it might be a comet. The planet lies more than twice as far from the Sun as Saturn, at nearly 3 billion km (1.8 billion mi.) away.

Uranus is unusual among the planets because its axis is tilted (at 98°) to roughly the plane of its orbit. This means that it spins on its side as it travels through space. Most of the other planets are more or less upright.

We cannot see Uranus with the naked eye, because it is too far away. And even a powerful telescope shows it only as a bluish-green disk. It has this color because its atmosphere contains methane. Methane absorbs the red wavelengths from white light, leaving blue-green. No features can be seen on the disk, even when it is viewed closely by space probes.

Like Jupiter, Uranus has faint rings around it that are not visible from Earth. But the *Voyager 2* probe saw them. There are at least 10 narrow rings, consisting of large chunks of dark rock. The thickest ring, the epsilon, is up to 100 km (60 mi.) wide. Rocks in the ring appear

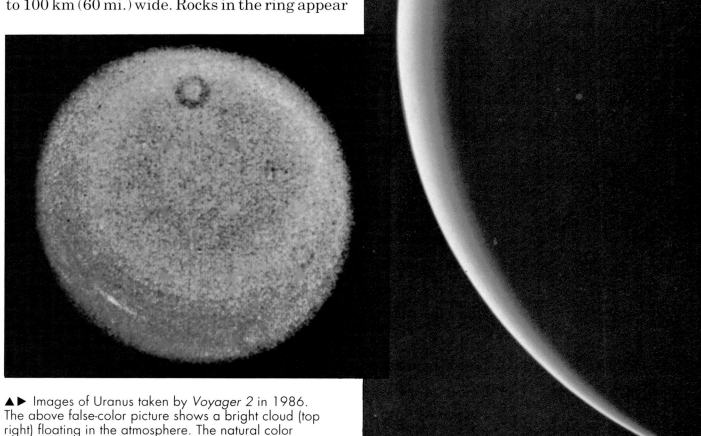

▲▶ Images of Uranus taken by *Voyager 2* in 1986. The above false-color picture shows a bright cloud (top right) floating in the atmosphere. The natural color picture (right) shows Uranus as a bright crescent.

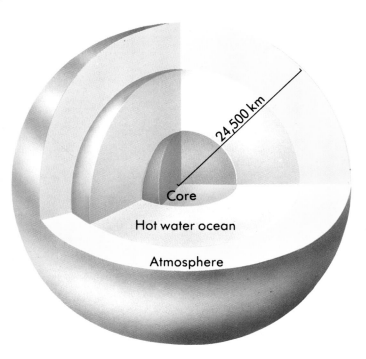

▲ Uranus has a heavy core of iron and rock. On top is an ocean of hot water and ammonia about 8,000 km (5,000 mi.) deep. The atmosphere is made up of hydrogen, helium, and methane.

▼ Wisps of cloud scurry across the atmosphere of Neptune in this image sent to Earth by *Voyager 2* in 1989. The red edge to the disk is a false-color effect showing haze. The dark spot is probably a storm.

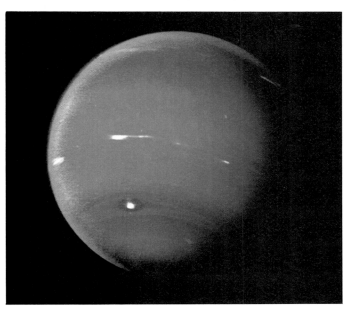

to be kept in place by a pair of tiny shepherd moons. At least 13 more moons circle the planet.

Twin Neptune

After Uranus was discovered, astronomers found that its orbit was not regular. And this made them think that there must be another planet farther out that was affecting Uranus by its gravity. In 1846 that planet was finally discovered by the German astronomer Johann Gottfried Galle. It was named Neptune.

Neptune is the fourth-largest planet, and is only a little smaller than Uranus. For most of the 165 years it takes to go once round its orbit, Neptune is the second-farthest planet from the Sun. But until 1999 it will be the farthest, because Pluto is currently traveling inside Neptune's orbit.

Astronomers can see only a few details of the planet through a telescope. But images sent back to Earth by *Voyager 2* have revealed a wealth of information. The overall color of the planet is deep blue, because of traces of methane in the atmosphere, which consists mainly of hydrogen and helium. It has a much more active weather system than Uranus.

Pluto and Charon

The ninth planet, Pluto, was discovered in 1930 by the American astronomer Clyde Tombaugh. For most of the 248 years it takes Pluto to orbit the Sun, it is the farthest planet. But at present it is traveling inside Neptune's orbit. Pluto is by far the smallest planet, smaller even than the Moon. It is probably made up mainly of rock and ice and has an atmosphere largely of methane. This image was taken by the Hubble Space Telescope. Pluto has one moon, called Charon, which is very large for a moon at half the diameter of its parent planet.

Debris of the Solar System

As well as planets and their moons, the Solar System contains many much smaller bodies. Those called asteroids circle the Sun in a broad belt, but they are too far away from Earth and too tiny to be seen with the naked eye. Other bodies make a more spectacular sight, sprouting a flaming head and a long glowing tail. We see them as comets. Smaller bits of rocky matter get trapped by Earth's gravity and plunge through the atmosphere, where they burn up. They leave fiery trails in their wake, which we call meteors, or shooting stars.

SPOT FACTS

• As many as 100 million meteorite particles burn up in the Earth's atmosphere every day.

• Every year the Earth gains up to 5 million metric tons in weight from meteorite particles that fall to the surface as dust.

• The massive meteorite that gouged out the Arizona Meteor Crater had the destructive power of a hydrogen bomb.

• Halley's comet has been observed on each of its returns since 87 BC.

• The nucleus of Halley's comet is about 15 km (9 mi.) long. It is velvety black and shaped like a potato.

• The tail of the very bright "daylight comet" of 1843 stretched 330 million km (200 million mi.) across the sky.

• Some asteroids discovered well beyond the orbit of Saturn are believed to be part of a new asteroid zone, called the Kuiper Belt, which extends beyond the orbit of Pluto.

• An asteroid called Ida, which was photographed in a closeup by the space probe *Galileo* on its way to Jupiter, was found to have a small moonlet accompanying it. The moonlet has been given the name Dactyl.

• When the Earth passes through dust from the tail of a comet we see a shower of meteors. This happens on regular dates, such as August 12 (Perseid meteors) and December 13 (Geminid meteors).

• The extinction of the dinosaurs 65 million years ago is widely believed to have been due to a giant asteroid impact, near the Gulf of Mexico.

• Many comets are discovered by amateur astronomers, who are rewarded by having the comet named after them.

• Comets originate in a huge, distant cloud surrounding the Sun known as the Oort Cloud. There are probably billions of comets in the Oort Cloud, which may stretch halfway to the nearest star.

• The gravitational pull of the giant planet Jupiter often alters the orbits of comets, either capturing comets on their way in to the Sun or flinging them out of the Solar System altogether.

• Despite their appearance, the bodies of comets can be very fragile. They frequently break up, never to be seen again.

• Although the tails of comets contain cyanogen, a gas linked to cyanide, the amount is so small that the Earth can pass right through the tail of a comet without any ill effects.

• In 1833, 1866, and 1899 the Earth passed through a swarm of meteors called the Leonids, producing hundreds of shooting stars a minute. A brief burst was seen in 1966.

• Most meteorites fall harmlessly into the sea. No one has been killed by a meteorite fall in recent times, though they have hit cars and houses.

• Some astronomers believe that there may be large reserves of iron and nickel on asteroids, and that in the future these may prove valuable resources.

Asteroids and comets

Many rocky, icy chunks were left over after the formation of the planets 5 billion years ago. The biggest collection of pieces ended up circling the Sun in a broad band between the orbits of Mars and Jupiter. These are the asteroids, or minor planets. Several thousand asteroids have been discovered, ranging in diameter from 1 km (about 0.5 mi.) to 1,000 km (600 mi.).

The asteroids are rocky bodies; the Sun long ago evaporated any ice they contained. But many lumps of matter left over since the birth of the Solar System are still in their original icy state. Most of them stay in the frozen depths of the Solar System and remain invisible. From time to time, however, some of them wander into the inner Solar System. The Sun evaporates the ice to form a gas cloud around them, and they become visible as comets.

Comets are among the most spectacular of all heavenly bodies. The brightest ones can grow a tail that stretches halfway across the sky. Some of them (in 1882 and 1910, for example) are visible in broad daylight. The coming of most comets cannot be predicted even today. In the past people were frightened by the appearance of a comet in the sky. They thought it brought bad luck or signaled disaster.

▲ This pitted lump of rock is the tiny Martian moon Phobos, which is less than 30 km (20 mi.) across. Probably Phobos was once an asteroid, circling at the edge of the asteroid belt. Some time long ago it wandered too close to Mars and became a satellite when it was captured by the planet's gravity.

◄ Halley's comet traveling against a starry background in April 1986. At that time it was getting near its closest approach to Earth, some 63 million km (39 million mi.). The comet was best observed that year in the Southern Hemisphere. Because the comet enters our region of the Solar System only once every 76 years, it will not be seen again until the year 2061. The telescope through which this picture was taken was trained on the comet during a 1-hour exposure. This caused the stars in the background to make trails. The comet was named after the British astronomer Edmond Halley. He was the first to recognize that records of past comets spotted at regular 76-year intervals, and having the same orbit, in fact referred to the same comet. The records he studied were dated 1531, 1607, and 1682, and he deduced that the comet would reappear in 1758, which it did.

Sizes and orbits

Most asteroids orbit the Sun in a band, or belt, which is about 150 million km (90 million mi.) wide. The inner edge of the belt lies about 300 million km (nearly 200 million mi.) from the Sun. The biggest asteroid, Ceres, is about 1,000 km (600 mi.) across. Only about 200 of the others are larger than 100 km (60 mi.) in diameter. The larger ones, such as Ceres, Pallas (600 km/370 mi. across), and Vesta (550 km/340 mi. across), appear to be ball-shaped. Eros, a smaller one, is a chunk of rock about 35 km by 15 km by 7 km (22 mi. by 10 mi. by 4 mi.).

Eros is also unusual in that it does not orbit in the main asteroid belt, but much closer in. Occasionally it comes within 25 million km (15

million mi.) of the Earth, as happened in 1975. In March 1989 an asteroid designated 1989FC came within 750,000 km (470,000 mi.). In cosmic terms, this was a very near miss.

The comet's tail
A comet is often described as a dirty "snowball" because it is made up of a mixture of dust and ice. We do not usually discover comets until they come within the orbit of Jupiter. Then the Sun's heat starts to evaporate the ice. This releases gas and dust to form a cloud, or coma, around the solid core, or nucleus.

As the comet gets closer to the Sun, the solar wind begins to affect it. The wind "blows" gas

Asteroid and comet orbits

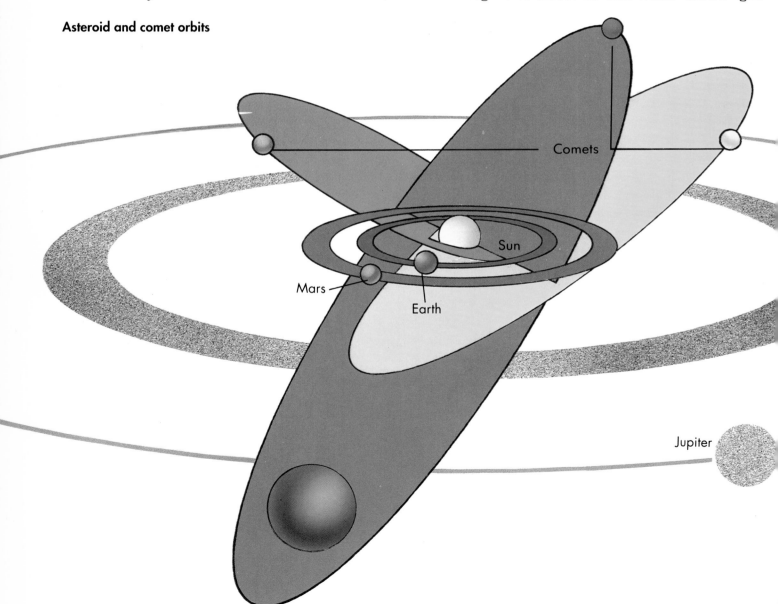

Comets

Sun

Mars

Earth

Jupiter

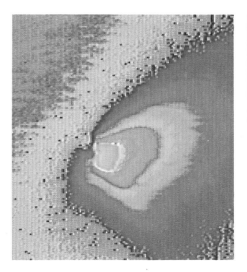

▲ A close-up image of Halley's comet, sent back by the European space probe *Giotto* in March 1986.

▶ The same comet, as photographed through a telescope in Australia at about the same time.

Asteroid belt

and dust particles away from the head of the comet into a long tail. The tail therefore always points away from the Sun. So when the comet swings round the Sun and starts receding, it travels tail-first. Gradually the tail shrinks and the comet fades. By the time it reaches Jupiter's orbit, it has all but disappeared from view.

Most bright comets appear suddenly, blaze for several weeks, and then disappear. It is nearly impossible to predict when they will return to Earthly skies. The arrival of some comets, however, is as regular as clockwork. Their orbits are accurately known; so are their periods – the time they take to travel around their orbit. The best known of these "periodic comets" is Halley's comet, with a period of about 76 years.

The comet's tail

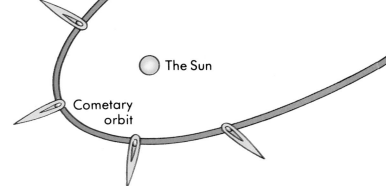

The Sun

Cometary orbit

▲ Most comets consist of a mixture of dust and ice. As they get near the Sun, the ice melts to form a reflecting tail that always points away from the Sun.

Meteors and meteorites

Day and night the Earth is bombarded with bits of rock from outer space. They are called meteoroids, and they vary in size from microscopic specks to huge boulders perhaps hundreds of feet in diameter. Almost certainly they are pieces of asteroids that have been broken off during collisions.

Meteoroid particles travel in orbit around the Sun, just like the Earth does. When they come near the Earth, they are attracted to it by gravity. They plunge into the upper atmosphere traveling at speeds up to 70 km/s (over 40 mps). Friction with the air heats up the particles and makes them glow white hot. In the night sky we see them as the fiery streaks we call meteors or shooting stars.

Most meteoroid particles are so small that they burn up completely. Others turn to dust and eventually settle on the ground. The biggest ones survive their fiery passage and fall to Earth as meteorites. If they are really huge they can create huge craters. Examples are Meteor Crater in the Arizona desert in the United States, and the Henbury Craters in New South Wales, Australia.

There are two main types of meteorites, stony and iron. There is also a somewhat rarer intermediate type, the stony-iron. Stony meteorites

▶ Meteor Crater in Arizona measures 1,265 m (4,150 ft.) across and is 175 m (574 ft.) deep. Today, only small pieces remain of the massive body that gouged out the crater some 25,000 years ago.

Flattened forest

◀ Trees were felled like bowlingpins by the blast from an explosion that occurred near the Podkamennaia Tunguska River in Siberia in June 1908. At first, astronomers thought that the explosion was caused by the impact of a huge meteorite. But they now think it was a small comet. They believe that the comet burst into the Earth's atmosphere and that its nucleus vaporized with explosive force several miles above the ground.

are made up mainly of silicates, like many stones on Earth. Many stony meteorites contain tiny rounded grains and are called chondrites. Some, called carbonaceous chondrites, are rich in carbon compounds.

Iron meteorites are made up mainly of iron and nickel, together with a little cobalt. When they are cut, polished, and etched with acid, a triangular crystal pattern shows up which is unique to meteorites.

No one is certain whether the pebble-sized objects called tektites come from outer space or not. They have a different composition from meteorites, and are similar to volcanic glass. They are found mainly in four areas, in North America, Australia, the Czech Republic and Slovakia, and the Ivory Coast in Africa.

Missiles from outer space

There are three kinds of meteorites: iron meteorites (1), called siderites; stony ones (2), called aeriolites; and stony-iron ones, called siderolites. Round-grained aeriolites containing traces of carbon are also known as carbonaceous chondrites. Tektites (3), possibly also from outer space, are glassy pebbles.

Myriads of moons

All the planets except Mercury and Venus have smaller bodies orbiting around them. These satellites, or moons, vary in size from lumps of rock a few tens of miles across to bodies bigger than Mercury. The giant outer planets have the most moons – Saturn has more than 20. All these tiny worlds are different from our own Moon, and from each other. Some are rugged and heavily cratered; others are as smooth as the ice that covers them. Some shine brightly, reflecting sunlight; others are dull and dark. Most are dead worlds, but at least one, Jupiter's Io, is alive with erupting volcanoes.

SPOT FACTS

• Earth's Moon is the sixth-largest satellite in the Solar System, after Ganymede, Callisto, and Io (orbiting Jupiter); Titan (orbiting Saturn); and Triton (orbiting Neptune).

• The largest crater visible on the Moon is Bailly, a "walled plain" nearly 300 km (200 mi.) across.

• A new mineral discovered in Moon rocks has been called armalcolite, for the three *Apollo 11* astronauts who took part in the first Moon landing – Armstrong, Aldrin, and Collins.

• The oldest Moon rock brought back by the *Apollo* astronauts is about 4.6 billion years old, the same age as the Earth.

• The Full Moon looks bigger when low in the sky than when high up, but this is just an optical illusion caused by having horizon features to judge its size.

• The only time when a true New Moon can be seen is during a total eclipse of the Sun. The earliest crescent Moon that we can see is usually at least 18 hours after true New Moon.

• Although there is no air on the Moon, a form of weathering does take place by the constant bombardment of particles from the solar wind. This wears down the outlines of the older craters.

• Jupiter's moon Io is the only body in the Solar System besides Earth where there are known to be active volcanoes.

• The interior of Io is hot because it is under constant strain from the gravitational pull of Jupiter.

• If Jupiter's moon Europa were reduced to the size of a pool ball, its surface features would be no thicker than a layer of ink.

• Callisto, another of Jupiter's moons, bears the scar of a giant asteroid impact in the distant past; a series of circular cracks cover half its diameter.

• Saturn's largest satellite, Titan, is believed to have oceans of methane and ethane beneath its cloud layers. There may even be primitive life forms there.

• Some of Saturn's smaller satellites seem to appear and disappear, so scientists believe that they are more likely to be clumps of debris rather than true satellites.

• Iapetus, a satellite of Saturn, is a two-tone satellite. One half is sooty, while the other half is icy and much brighter. The reason for this is unknown.

• Miranda, a strange satellite of Uranus, has ice cliffs some 20 km (12 mi.) high. This is greater than the difference between the Earth's highest mountain and its deepest ocean.

• Neptune's largest satellite, Triton, is the only large moon to orbit its planet in the opposite direction to the planet's rotation. This is seen as evidence that it was captured some time after Neptune's formation.

• Charon is so large compared with Pluto that the pair are often considered to be a double planet rather than a planet and moon.

The Moon

Phases of the Moon

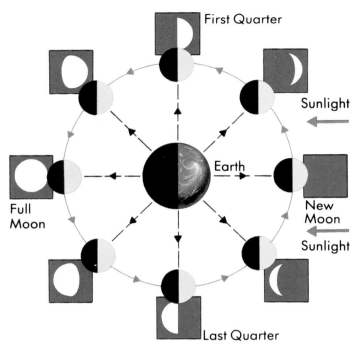

First Quarter

Sunlight

Earth

Full Moon

New Moon

Sunlight

Last Quarter

▲ The Moon gives off no light of its own, but shines by reflecting sunlight. As it travels around the Earth, we see more or less of the surface lit up. It all depends on the position of the Moon in relation to the Sun. From the Earth it appears that the Moon is changing shape. We call the changing shapes of the Moon its phases. It takes 29½ days to go through all the phases. When the Moon lies between the Earth and the Sun, it shows us a dark face (New Moon). Later, a crescent appears and grows in size. A week after New Moon, half the Moon is lit up (First Quarter). The visible area increases, in a "gibbous" phase, until two weeks after New Moon the whole face is lit up (Full Moon). Then it gradually shrinks again through gibbous, half Moon (Last Quarter), and crescent, until it disappears at the next New Moon.

▶ The structure of the Moon resembles that of a small planet. It probably has a small core, surrounded by a partly molten zone. Above is a solid mantle, covered by a thin crust.

The Earth has just one natural satellite, the Moon. We know more about it than about any other moon because it is our closest neighbor. And astronauts have explored its surface.

The Moon lies at an average distance of 384,000 km (239,000 mi.) from Earth, and is more than 100 times closer than the nearest planet. Its diameter (3,476 km, or 2,160 mi.) is nearly one-quarter that of the Earth, which is large for a satellite.

Earth's gravity keeps the Moon circling in orbit around it, completing an orbit once every 27⅓ days. The Moon also spins on its axis once during this time. This results in the Moon always presenting the same face toward us. Lunar gravity is much less than the Earth's (one-sixth) because the Moon has much less mass. But it still affects the Earth by causing the rise and fall of the tides.

With its low gravity, the Moon has been unable to keep any significant atmosphere, so there is no weather of any kind. The Moon is a dead, silent world. There is a marked contrast between day and night. In the day temperatures soar to over 100°C (200°F), but at night they drop to as low as -150°C (-240°F).

The Moon's structure

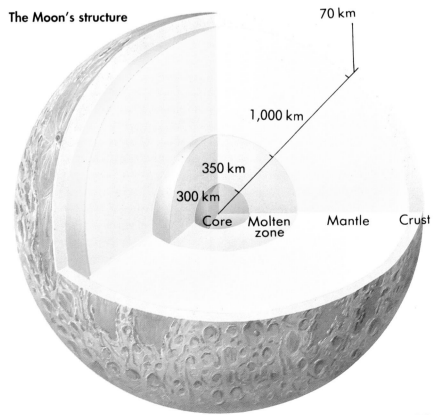

70 km

1,000 km

350 km

300 km

Core   Molten zone   Mantle   Crust

The lunar surface

These two photographs both show a full face of the Moon. But notice how different they are. The picture on the left shows the face of the Full Moon as we see it from Earth. Much of it is covered by the great lava plains we call seas. The picture above was taken from space by the *Apollo 11* astronauts. The right-hand part shows regions on the far side of the Moon that we can never see from the Earth. Notice that there are no large seas there, just rugged and cratered highlands.

Seas and highlands

Even with the naked eye we can see that the Moon's surface is made up of two main areas, dark and light. Through a telescope we can see that the dark areas are vast flat plains, while the light ones are rugged highlands.

The dark plains are vast sheets of lava. They were formed billions of years ago when massive meteorites slammed into the Moon and melted the rocks. Early astronomers thought these areas might be seas, and called them maria (Latin for "seas"). Some seas, such as the Sea of Crises, are circular and surrounded by mountains. Other seas merge together. The largest sea area is located in the northwest, where the Ocean of Storms, the Sea of Showers, and the Sea of Clouds merge together.

The light-colored highland areas of the Moon are part of the Moon's ancient crust. They are much more heavily cratered than the seas. Strangely, the far side of the Moon is almost entirely a highland region. It has only one sea of any size, the Sea of Moscow.

The highest mountain ranges on the Moon border the seas. The Sea of Showers is ringed by the lofty Lunar Alps, Caucasus, Apennine, and Carpathian ranges, in which some peaks rise to 6,000 m (20,000 ft.).

We now know what the lunar surface is like, thanks to the 12 *Apollo* astronauts who walked on the Moon. They brought back some 385 kg (850 lb.) of soil and rock. The dusty topsoil is made up of particles of rock smashed to pieces by meteorites. The main type of rock in the seas is a dark volcanic rock like basalt. The highlands are composed of a lighter volcanic rock. Everywhere there are breccias, rocks made up of a cemented mixture of rock chips.

▼ Man on the Moon. Harrison Schmitt examines a huge split boulder in the Taurus-Littrow Valley on the last *Apollo* mission, *Apollo 17*, in December 1972.

▶ The crater Eratosthenes, on the edge of the Sea of Showers, measures about 65 km (40 mi.) across. The central mountain peaks are typical of large lunar craters.

Moons of the planets

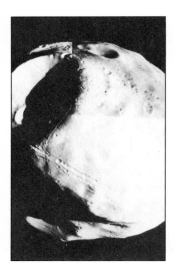 

◄ The two tiny moons of Mars are Phobos (left) and Deimós. Unlike most moons, they are odd-shaped lumps of rock, and both are covered in craters. One huge crater on Phobos measures 10 km (6 mi.) across.

Martian moons

The two moons of Mars are rocky bodies of irregular shape. They were probably asteroids that orbited close to Mars and were captured.

Even the largest moon, Phobos, is only about 28 km (17 mi.) across. It orbits about 6,000 km (nearly 4,000 mi.) above the surface and speeds around the planet in only about 7½ hours. This is much faster than Mars itself rotates (24⅔ hours). So from the Martian surface the moon would appear to move from west to east, the opposite way from usual. Deimos is about 16 km (10 mi.) across. It circles some 20,000 km (12,000 mi.) away in just over a day.

We cannot see any details of the other moons in the Solar System even through a telescope. They are too small and far away. Fortunately, space probes have visited many of the moons and sent back close-up photographs.

Most moons, like our own, orbit close to the plane of their parent planet's equator. And most travel in a nearly circular orbit in a counterclockwise direction, when viewed from the north of the Solar System.

Jupiter's Galilean quartet

Apart from our own Moon, the easiest moons to spot are the four biggest moons of Jupiter. We can see them easily using good binoculars. The Italian astronomer Galileo first spotted them in the winter of 1609/10, with his telescope. So they are called the Galilean satellites.

Active Io

Jupiter's large moon Io looks quite different from any other moon in the Solar System. It is a vivid orange-yellow, speckled with black (right of picture). When the *Voyager* space probes flew past the moon, they took close-up pictures of active volcanoes.

▲ Io is the third largest of Jupiter's four Galilean moons. But it is the most lively. Its active volcanoes eject matter to heights of 250 km (150 mi.) or more. The orange color is thought to be caused by the presence of sulfur.

► On Jupiter's moon Ganymede, dark regions alternate with paler grooved areas. The bright craters show where meteorites have fallen to the surface and exposed fresh patches of white ice.

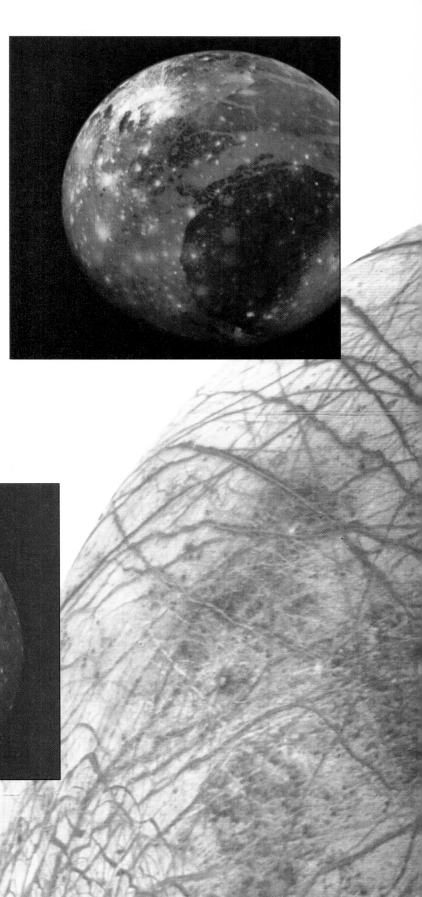

These four large moons are Io, Europa, Ganymede, and Callisto, in order of distance from the planet. With a diameter of 5,276 km (3,278 mi.), Ganymede is the biggest moon in the Solar System, bigger even than the planet Mercury. Callisto is of similar size. Both are made up of a mixture of rock and ice.

Europa and Io are about the size of the Moon and are made up mainly of rock. But they could not look more different. Europa has a very smooth icy surface, crisscrossed with a network of dark lines. These are probably fractures in the ice that have become filled with darker material from underneath. Io is a vivid orange-yellow and is volcanically active. All the other 12 or so Jovian moons are very much smaller. Four of them orbit closer to the planet than Io. Two other groups of four moons orbit very much farther out. The four most distant moons have retrograde (backward) and highly eccentric orbits. Astronomers think that they could well be captured asteroids.

▲ Callisto has an ancient crust, completely covered with craters.

► Europa has an icy crust, which is extremely smooth. There are few signs of any craters.

New discoveries

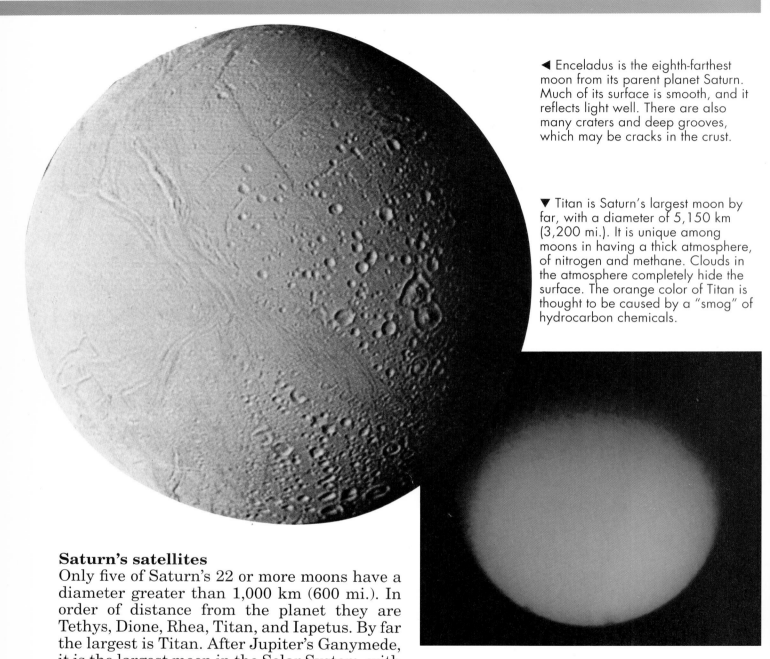

◄ Enceladus is the eighth-farthest moon from its parent planet Saturn. Much of its surface is smooth, and it reflects light well. There are also many craters and deep grooves, which may be cracks in the crust.

▼ Titan is Saturn's largest moon by far, with a diameter of 5,150 km (3,200 mi.). It is unique among moons in having a thick atmosphere, of nitrogen and methane. Clouds in the atmosphere completely hide the surface. The orange color of Titan is thought to be caused by a "smog" of hydrocarbon chemicals.

Saturn's satellites

Only five of Saturn's 22 or more moons have a diameter greater than 1,000 km (600 mi.). In order of distance from the planet they are Tethys, Dione, Rhea, Titan, and Iapetus. By far the largest is Titan. After Jupiter's Ganymede, it is the largest moon in the Solar System, with a diameter of 5,150 km (3,200 mi.). It is particularly interesting because it has an atmosphere which is denser than Earth's.

Most of the moons are made up of rock and ice and are heavily cratered. Mimas has an enormous crater, 130 km (80 mi.) across – a third of its diameter! The outermost moon, tiny Phoebe, is probably a captured asteroid.

The moons of Uranus

We can see only five of the 15 or so moons of Uranus through a telescope. Going out from the planet, they are Miranda, Ariel, Umbriel, Titania, and Oberon. They seem to be made up of rock and ice in about equal proportions. All are covered with craters, some of which show up bright where fresh ice has been thrown out.

Miranda is by far the most interesting moon. Its surface is a patchwork of totally different kinds of regions, with distinct boundaries between them. Rolling cratered plains suddenly give way to curious grooved regions. In the past Miranda may have collided with an asteroid and been smashed to pieces. Then gravity pulled the pieces back together again to re-form the moon.

▲ Miranda, the fifth-largest moon of Uranus, is irregularly shaped with a diameter of about 480 km (300 mi.). As well as the strange grooves, it has ice cliffs some 20 km (12 mi.) high.

▼ Pink "snow" covers most of the southern hemisphere of Neptune's moon Triton. The snow is a mixture of frozen methane and nitrogen.

Neptune's new moons

Viewed from Earth, Neptune appears to have only two moons. The largest and the one closest to the planet is Triton, which is only a little smaller than our own Moon. The other is Nereid, only about 170 km (110 mi.) across.

In 1989, *Voyager 2* discovered a further six moons. One is rather bigger than Nereid and circles close to the planet. Four others are located inside Neptune's faint ring system.

Voyager's close-up view of Triton showed a fascinating world tinged with pink. It has a faint atmosphere of nitrogen. It is also the coldest known place in the Solar System, at a temperature of some -240°C (-400°F).

Pluto's Charon

The American astronomer James Christy discovered Pluto's icy moon Charon in 1978. Its diameter (about 1,190 km, or 740 mi.) is no less than half that of Pluto! It orbits at a distance of only about 20,000 km (12,000 mi.). It makes one revolution about every six Earth-days, the same time it takes Pluto to spin on its axis.

Units of measurement

Units of measurement

This encyclopedia gives measurements in metric units, which are commonly used in science. Approximate equivalents in traditional American units, sometimes called U.S. customary units, are also given in the text, in parentheses.

Some common metric and U.S. units

Here are some equivalents, accurate to parts per million. For many practical purposes rougher equivalents may be adequate, especially when the quantity being converted from one system to the other is known with an accuracy of just one or two digits. Equivalents marked with an asterisk (*) are exact.

Volume
1 cubic centimeter = 0.0610237 cubic inch
1 cubic meter = 35.3147 cubic feet
1 cubic meter = 1.30795 cubic yards
1 cubic kilometer = 0.239913 cubic mile

1 cubic inch = 16.3871 cubic centimeters
1 cubic foot = 0.0283168 cubic meter
1 cubic yard = 0.764555 cubic meter

Liquid measure
1 milliliter = 0.0338140 fluidounce
1 liter = 1.05669 quarts

1 fluidounce = 29.5735 milliliters
1 quart = 0.946353 liter

Mass and weight
1 gram = 0.0352740 ounce
1 kilogram = 2.20462 pounds
1 metric ton (tonne) = 1.10231 tons

1 ounce = 28.3495 grams
1 pound = 0.453592 kilogram
1 short ton = 0.907185 metric ton (tonne)

Length
1 millimeter = 0.0393701 inch
1 centimeter = 0.393701 inch
1 meter = 3.28084 feet
1 meter = 1.09361 yards
1 kilometer = 0.621371 mile

1 inch = 2.54* centimeters
1 foot = 0.3048* meter
1 yard = 0.9144* meter
1 mile = 1.60934 kilometers

Area
1 square centimeter = 0.155000 square inch
1 square meter = 10.7639 square feet
1 square meter = 1.19599 square yards
1 square kilometer = 0.386102 square mile

1 square inch = 6.4516* square centimeters
1 square foot = 0.0929030 square meter
1 square yard = 0.836127 square meter
1 square mile = 2.58999 square kilometers

1 hectare = 2.47105 acres
1 acre = 0.404686 hectare

Temperature conversions

To convert temperatures in degrees Celsius to temperatures in degrees Fahrenheit, or vice versa, use these formulas:

Celsius Temperature = (Fahrenheit Temperature − 32) × 5/9
Fahrenheit Temperature = (Celsius Temperature × 9/5) + 32

Numbers and abbreviations

Numbers

Scientific measurements sometimes involve extremely large numbers. Scientists often express large numbers in a concise "exponential" form using powers of 10. The number one billion, or 1,000,000,000, if written in this form, would be 10^9; three billion, or 3,000,000,000, would be 3×10^9. The "exponent" 9 tells you that there are nine zeros following the 3. More complicated numbers can be written in this way by using decimals; for example, 3.756×10^9 is the same as 3,756,000,000.

Very small numbers – numbers close to zero – can be written in exponential form with a minus sign on the exponent. For example, one-billionth, which is 1/1,000,000,000 or 0.000000001, would be 10^{-9}. Here, the 9 in the exponent –9 tells you that, in the decimal form of the number, the 1 is in the ninth place to the right of the decimal point. Three-billionths, or 3/1,000,000,000, would be 3×10^{-9}; accordingly, 3.756×10^{-9} would mean 0.000000003756 (or 3.756/1,000,000,000).

Here are the American names of some powers of ten, and how they are written in numerals:

1 million (10^6)	1,000,000
1 billion (10^9)	1,000,000,000
1 trillion (10^{12})	1,000,000,000,000
1 quadrillion (10^{15})	1,000,000,000,000,000
1 quintillion (10^{18})	1,000,000,000,000,000,000
1 sextillion (10^{21})	1,000,000,000,000,000,000,000
1 septillion (10^{24})	1,000,000,000,000,000,000,000,000

Principal abbreviations used in the encyclopedia

°C	degrees Celsius
cc	cubic centimeter
cm	centimeter
cu.	cubic
d	days
°F	degrees Fahrenheit
fl. oz.	fluidounce
fps	feet per second
ft.	foot
g	gram
h	hour
Hz	hertz
in.	inch
K	kelvin (degree temperature)

kg	kilogram
l	liter
lb.	pound
m	meter
mi.	mile
ml	milliliter
mm	millimeter
mph	miles per hour
mps	miles per second
mya	millions of years ago
N	north
oz.	ounce
qt.	quart
s	second
S	south
sq.	square
V	volt
y	year
yd.	yard

Glossary

absolute magnitude A measure of the true brightness of a star. It is the apparent magnitude a star would have at a distance of 10 parsecs, or 32.6 light-years.

active galaxy One that has an exceptionally high energy output at light, radio, or other wavelengths.

albedo A measure of a body's ability to reflect light. It is the ratio of the amount of light reflected to the amount received. The Moon's albedo is only 0.07.

annular eclipse An eclipse of the Sun, during which the face of the Moon does not quite cover all of the Sun, leaving a bright ring (annulus).

aphelion The point in its orbit where a planet is farthest away from the Sun.

apogee The point in its orbit where the Moon or an artificial satellite is farthest away from the Earth.

apparent magnitude A measure of the brightness of a star as it appears to us.

asteroid A small rocky body that orbits (usually) between the orbits of Mars and Jupiter. Also called minor planet.

astrology A pseudoscience founded on the belief that human affairs are influenced by the positions of the heavenly bodies.

astronomical unit (AU) The average distance between the Earth and the Sun, 149,600,000 km (93 million mi.).

astronomy The scientific study of the heavens.

Big Bang The explosion that is thought to have created the Universe about 15 billion years ago.

Big Crunch An event that could happen at the end of a closed Universe, when all energy and matter would come together in a reverse of the Big Bang.

binary A double star in which the two component stars revolve around each other. A visual binary is one in which the two stars are far enough apart to be seen separately in a telescope. A spectroscopic binary is one in which the stars are so close together that they can be identified only from a spectrum.

black hole A superdense body whose gravity is so intense that not even light can escape from it.

blue shift The shift in the spectral lines of starlight from their normal position toward the blue end of the spectrum. The shift is caused by the star traveling towards us, in effect shortening the wavelength of the light.

brightness See **magnitude.**

celestial equator An imaginary line where the plane of the Earth's Equator meets the celestial sphere.

celestial sphere An imaginary sphere around the Earth to which the stars seem to be fixed.

Cepheids A class of variable stars which vary in brightness as regular as clockwork. Their period of variation is related to their absolute magnitude, or true brightness.

chromosphere ("color sphere") The first layer of the Sun's atmosphere. It has a reddish color and can be seen only during a total eclipse of the Sun.

coma The cloud that forms the head of a comet, made up of dust and gas.

comet A primitive member of the Solar System, consisting of rock, ice, and dust. It starts to shine when it approaches the Sun.

conjunction The apparent close approach of heavenly bodies to each other.

constellation A pattern made by a group of bright stars. Although they appear as a group in our line of sight, they lie at widely different distances in space.

corona The pearly white halo, or crown, visible around the Sun during a total solar eclipse.

cosmic rays Radiation coming from outer space, in part from the Sun as the solar wind.

cosmic year The time it takes the Sun to travel once around the Galaxy, about 225 million Earth years.

cosmos Another word for the Universe.

crater A hole in the surface of a planet or a moon, usually made by the impact of meteorites.

crater rays Bright rays that fan out from some of the large lunar craters, such as Tycho. They are probably strings of glassy material thrown out when the crater was formed.

declination A star's "latitude" on the celestial sphere. It is measured in degrees north (+) or south (-) of the celestial equator.

double star A pair of stars that appear close together in the heavens. An optical double is a pair of stars that appear close only because they are in the same line of sight, but actually lie at very different distances. A binary is a system of two stars that revolve around each other.

eclipse What happens when one heavenly body passes in front of another and blots out its light. An eclipse of the Sun occurs when the Moon blots out the Sun's light. An eclipse of the Moon occurs when the Moon moves into the Earth's shadow in space.

eclipsing binary A binary star system in which the two stars orbit in a plane which takes one in front of the other as we see them from Earth. As each star is eclipsed by the other, the overall brightness of the system decreases momentarily.

ecliptic The apparent path of the Sun each year around the celestial sphere.

electromagnetic radiation The kind of radiation given out by glowing bodies like stars. It consists of electric and magnetic waves, or vibrations. Light, radio waves, ultraviolet rays, X-rays, and gamma rays are all kinds of electromagnetic radiation; they differ in their wavelength.

encounter The meeting in space between a space probe and a planet or a moon.

equator An imaginary line around a planet or moon in the plane of its rotation, such as the Earth's Equator. It is equidistant from the poles.

equinox A time when the lengths of the day and night are equal. This happens twice a year when the Sun crosses the celestial equator. On about March 21 the Sun is moving north (vernal, or spring, equinox). On about September 23, it is moving south (autumnal equinox). The equinoxes mark where the ecliptic intersects the celestial equator.

evening star A name for the planet Venus when it shines brightly in the western sky at sunset.

expanding Universe Most of the galaxies seem to be rushing away from us, making it appear that the whole Universe is expanding. Most astronomers now believe that it is expanding.

Fraunhofer lines The dark lines in the spectrum of the Sun, first explained by the German astronomer Joseph von Fraunhofer.

galaxy An island of stars in space, which is usually elliptical or spiral in shape. The galaxy to which our Sun belongs is called the Galaxy or Milky Way.

gibbous The phase of the Moon (or other body) when more than half, but not all, of its face is lit up.

globular cluster A group of stars containing up to hundreds of thousands of members, clustered together into a globe shape. Several hundred such clusters circle around the centers of galaxies.

gravitation, or gravity The force of attraction that exists between any bodies. It is one of the great forces of the Universe.

Great Red Spot A huge oval region observed on the disk of Jupiter for centuries. It is thought to be a massive storm center.

helium The second most abundant element in the Universe, after hydrogen. It was first identified in the Sun, and was named after the Greek word for Sun, *helios*.

Hertzsprung-Russell diagram A kind of graph on which data about the stars are plotted. The stars' true brightness (absolute magnitude or luminosity) is plotted on the vertical, or y, axis; their spectral class or temperature is plotted on the horizontal, or x, axis.

Hubble classification The method used by an American astronomer to classify the galaxies – into elliptical galaxies, spirals, and barred spirals.

hydrogen The most plentiful element in the Universe and also the simplest. Its atoms most often consist of a single proton and a single electron.

interplanetary Between the planets.

interstellar Between the stars.

interstellar matter Gas and dust found in the space between the stars.

librations Slight irregularities in the Moon's motion around the Earth which allow us to see more than half of its surface.

light-year The common unit for measuring distances in astronomy, being the distance light travels in a year. 1 light-year = 9,460,000,000,000 km, or about 10 trillion km (almost 6 trillion mi.).

luminosity A measure of the true brightness of a star. It is expressed in terms of the star's energy output, or more usually as a comparison with the luminosity of the Sun.

lunar Relating to the Moon.

magnetosphere A region in space around a heavenly body where its magnetic field exerts an influence.

magnitude A scale on which star brightness is measured. It is based on six levels of brightness, from the brightest stars in the sky (1st magnitude) to the ones just visible to the naked eye (6th magnitude).

main sequence The diagonal band on the Hertzsprung-Russell diagram on which most stars lie. It represents the time of a star's life when it is stable and shining steadily.

mare A dark dusty plain on the Moon. It is the Latin word for sea, and its plural is *maria*. Early astronomers thought that the dark areas on the Moon could be seas.

meridian A great circle on the celestial sphere passing through the zenith and the celestial poles.

meteor The streak of light created when a piece of rock from outer space plunges through the Earth's atmosphere and is heated white-hot by air friction.

meteorite A piece of rock from outer space that survives fiery passage through the atmosphere and falls to the ground.

meteoroid A particle or piece of rock in orbit around the Sun.

Milky Way A fuzzy band of light arcing across the heavens. It represents a cross section of our

Galaxy.

morning star The planet Venus when it shines brightly in the dawn sky.

nadir A point on the celestial sphere directly below an observer.

nebula A cloud of dust and gas in space. It may shine on its own account or reflect light (bright nebula), or it may blot out the light from stars behind it (dark nebula).

neutron star A star made up of neutrons. It is incredibly dense.

North Star A bright star close to the northern celestial pole. It is also called the polestar and Polaris. There is no convenient similar star in the Southern Hemisphere.

Northern Lights The display of aurora that takes place at high latitudes in the Northern Hemisphere; the Aurora Borealis.

nova A star that suddenly increases dramatically in magnitude.

nuclear energy Energy given off when the nucleus of an atom splits (a process called fission) or combines with the nuclei of other atoms (a process called fusion). The fusion of hydrogen atoms produces the energy that makes the stars shine.

nucleus, comet The part of a comet that contains most of its mass, consisting of rock, dust, and ice. See also **coma.**

open cluster A loose group of stars containing typically a few hundred members.

opposition The position of a planet when it lies directly behind the Earth as seen from the Sun. It is an excellent time for observation.

orbit The path in space of one body around another, such as a moon around a planet.

parallax The apparent shift in a nearby object against a distant background when viewed from a different angle. This principle is used to measure the distance to nearby stars.

parsec A unit used to measure distance in astronomy. It is the distance at which a star would show a parallax shift of 1 second of arc (1/3600 of a degree). It is equal to 3.26 light-years.

partial eclipse An eclipse of the Sun in which the disk of the Moon covers only part of the Sun's disk.

perigee The point in its orbit where a Moon or satellite is closest to Earth.

perihelion The point in its orbit where a planet or comet is closest to the Sun.

periodic comet One that orbits the Sun and appears in the heavens at regular intervals, such as Halley's comet.

phase The different shape presented by the Moon, Mercury, or Venus, as more or less of it is lit by the Sun.

photosphere ("light sphere") The bright visible surface of the Sun.

planet A body that revolves around a star, particularly a body that revolves around the Sun. Planets generate no light of their own, but shine by reflecting sunlight. See also **Solar System**.

planetary nebula A mass of glowing gas often seen as a disk shape, rather like a planet's disk. It is a shell of gas puffed out by a central star as it dies.

poles Points on a heavenly body on the axis of rotation, the imaginary line around which the body rotates. The Earth's poles are the North Pole and the South Pole.

precession The slow movement of the celestial poles on the celestial sphere, due to the Earth "wobbling" on its axis. Because of precession the equinoxes change by one month every 2,000 years.

probe A spacecraft sent deep into space to visit planets, moons, and other heavenly bodies.

prominence A fountain of flaming gas that shoots high above the Sun's surface.

proper motion The movement of a star across the line of sight; detectable for some nearby stars.

pulsar A neutron star that emits rapid pulses of radiation as it rotates.

quasar Or quasi-stellar object (QSO); a body that looks like a star, but is very remote, and has the energy output of hundreds or thousands of galaxies.

radio astronomy The branch of astronomy that studies heavenly bodies by the radio waves they give out. It uses radio telescopes to gather the radio waves.

radio galaxy A galaxy that pours out much more energy than usual at radio wavelengths.

red giant A very large star which gives off mainly red light. It represents a late stage in the life of a star like the Sun.

red shift The shift in the spectral lines of starlight from their normal position toward the red end of the spectrum.

retrograde motion Motion in the opposite direction from usual.

right ascension Star "longitude." It is measured in time units eastwards along the celestial equator, from the vernal equinox.

satellite A body that revolves around another; a moon. Manufactured moons launched into orbit around the Earth are usually termed satellites, but are properly described as artificial satellites.

sea, lunar. See **mare.**

seasons The regular changes in weather conditions that occur on Earth during the year. They come about because of the tilt of the Earth's axis in space. Seasonal changes take place on other planets as well, for example, Mars.

shepherd moons Small moons found near a planet's ring system, which seem to help keep the ring particles in place.

sidereal time Time measured by the Earth's rotation with respect to the stars.

solar Relating to the Sun.

Solar System The family of heavenly bodies that includes the Sun and all the bodies that circle around it in space, such as planets and their moons, asteroids, and comets.

solar wind The constant stream of charged particles given off by the Sun, mainly protons and electrons.

solstice A time of the year when the Sun reaches the farthest point north or south on the celestial sphere. This occurs on about June 21 (summer solstice) and December 22 (winter solstice) each year.

Southern Lights The display of aurora that takes place at high altitudes in the Southern Hemisphere; the Aurora Australis.

spectroscope An instrument used to split up starlight into a spectrum and to study that spectrum. A spectrograph takes photographs of the spectrum.

spectrum A display of the range of wavelengths of radiation given off by a star.

star A gaseous body that produces its own energy as light, heat, and other radiation.

stellar Relating to the stars.

sunspot A darker and cooler region on the Sun. Sunspots come and go according to a regular cycle (called the sunspot, or solar, cycle) of about 11 years.

supergiant A very large, massive star with a diameter typically several hundred times that of the Sun.

supernova A star that increases in brightness millions of times as it blasts itself apart.

synchronous rotation A state in which a moon spins on its axis in exactly the same time it orbits its parent planet. This results in the moon keeping the same face pointing toward the planet. The Moon has a synchronous rotation in its orbit around the Earth.

telescope The main instrument used by astronomers to study the stars. Refracting telescopes use lenses to gather starlight; reflecting telescopes use curved mirrors to gather the light.

terrestrial Relating to the Earth.

tides The twice-daily rise and fall of the oceans brought about by the Moon's gravitational pull.

total eclipse An eclipse of the Sun in which the disk of the Moon masks all of the Sun's disk.

transit The passage of a smaller heavenly body across another, such as the planets Mercury and Venus across the Sun's disk.

Universe All that exists: Earth, Sun, stars, and even space itself.

Van Allen belts Doughnut-shaped bands of intense radiation that circle the Earth. They are formed by charged particles from the solar wind being trapped by the Earth's magnetic field.

variable star One whose brightness varies, often in a regular manner.

white dwarf A hot, very dense, planet-sized star near the end of its life.

zodiac A band around the celestial sphere centered on the ecliptic, in which the Sun, Moon, and major planets can always be found. It is divided into 12 parts, or signs, each corresponding to a different constellation.

Index

Page numbers in *italics* refer to pictures. Users of this Index should note that explanations of many scientific terms can be found in the Glossary.

Further reading

Atkinson, Stuart. *Astronomy*. Tulsa, OK: EDC Publishing, 1995.

Brewer, Duncan. *Comets, Asteroids, and Meteorites*. Tarrytown, NY: Marshall Cavendish, 1992.

Couper, H., and Nigel Henbest. *The Space Atlas: A Pictorial Guide to Our Universe*. Orlando, FL: Harcourt Brace & Co., 1992.

Erickson, Jon S. *Target Earth! Asteroid Collisions Past and Future*. Blue Ridge Summit, PA: TAB Books, 1991.

Estalella, Robert. *Galaxies*. Hauppauge, NY: Barron's Educational Series, 1994.

George, Michael. *Galaxies*. Mankato, MN: Creative Education, 1993.

George, Michael. *Stars*. Mankato, MN: Creative Education, 1992.

Graham, Ian. *Astronomy*. Chatham, NJ: Raintree Steck-Vaughn, 1995.

Gustafson, John R. *Planets, Moons, and Meteors*. Morristown, NJ: Silver Burdett Press, 1992.

Gustafson, John R. *Stars, Clusters, and Galaxies*. Morristown, NJ: Silver Burdett Press, 1993.

Mahy, Margaret. *The Greatest Show off Earth*. New York: Viking Children's Books, 1994.

Miotto, Enrico. *The Universe: Origin and Evolution*. Chatham, NJ: Raintree Steck-Vaughn, 1994.

Muirden, James. *Stars and Planets*. New York: Larousse Kingfisher Chambers, 1993.

Ridpath, Ian. *Atlas of Stars and Planets*. New York: Facts on File, 1993.

Stannard, Russell. *Our Universe: A Guide to What's Out There*. New York: Larousse Kingfisher Chambers, 1995.

Stephenson, Robert, and Roger Browne. *Exploring Earth in Space*. Chatham, NJ: Raintree Steck-Vaughn, 1992.

Stott, Carole. *Night Sky*. New York: Dorling Kindersley, 1994.

Stott, Carole (ed.). *Space Facts*. New York: Dorling Kindersley, 1995.